Food Additives, Nutrients & Supplements A-To-Z

Food Additives, Nutrients & Supplements A-To-Z

A SHOPPER'S GUIDE

Eileen Renders, N.D.

CLEAR LIGHT PUBLISHERS
SANTA FE

© 1999 Eileen Renders

Clear Light Publishers
823 Don Diego, Santa Fe, NM 87501
WEB: www.clearlightbooks.com

First Edition
10 9 8 7 6 5 4 3 2 1

Library of Congress Cataloging-in-Publication Data

Renders, Eileen, 1939-
    Food Additives, nutrients and supplements, A-to-Z : a shopper's
guide / Eileen Renders.
        p. cm.
    Includes index.
    ISBN 1-57416-008-7
    1. Food additives. 2. Nutrition.
TX553.A3R47   1998
664'.06—dc21   98-11905
                                                            CIP

This health information reference book is designed to provide current
general information on food additives, dietary components, and
dietary and therapeutic supplements, including herbs. The information contained in this book is not intended to be used as a substitute
for medical advice, diagnosis, or treatment.

Contents

Dedication

This book is dedicated to my grandchildren
—Joey, Ryan, Marissa, Erin, and Danielle—
in the hope that they and all our children may come
to understand the importance of maintaining
the most precious God-given gift, our health.

Acknowledgment

The author wishes to acknowledge the effort and teamwork of all who made this book possible and to express thanks and appreciation for their contribution. Special thanks are extended to Clear Light Publishers for their belief in and acceptance of this work. To the editors I will be forever grateful.

Highest gratitude must be given to God. The healing powers of nature that He has provided have attracted the attention and interest of past and present natural healing practitioners and will inspire healing professionals of the future.

Preface

This book was born out of a need to share information with others who are concerned with the complexities of health care issues that confront all of us today. It also was triggered by personal events that started me on a path of learning and study. Like many other alternative health care practitioners, I was initially motivated to choose this career because of personal health problems for which mainstream medicine had provided no solution. The alternative health options I encountered as I studied and learned remain my passion for each new day. And with the development of a career in Naturopathy came the immense personal gratification that comes from helping others.

There is nothing about my story that is unique. In fact, the frustrations I experienced have become commonplace, and doubtless many readers can relate to them. I'm offering a brief account of my own story in hopes that it will provide encouragement to others who have come to a point when traditional medicine offers no further hope.

As I began to search for alternative solutions, I repeatedly asked myself, "Could this disease have been prevented?" I soon realized that proper nutrition would help to fortify my body's resistance to disease. Healthier lifestyle habits might provide a simple path to maintaining an ideal weight, or assist my body in managing and reducing the stresses caused by everyday living. I sought a professional who might be qualified to guide and teach me the answers to the many questions I had, but I could find no one.

God works in mysterious ways, for suddenly it became clear to me that I needed to become that professional I had been seeking if I was to become the individual that I knew I wanted to be. Knowing that it wouldn't be easy, and that it would take patience and determination, I plunged ahead.

Because so many alternative health care practitioners became interested in such a career as a result of dealing with a personal health crisis, they are familiar with the personal transitions each of their clients must also make. Many clients come to the doctor of Naturopathy for help because they have become aware of their intolerance to drugs. Personally, I felt that since medications cannot "cure" and may only temporarily alleviate or mask the symptoms associated with many disorders, they were definitely not for me! However, avoiding medications is certainly not to be undertaken lightly, and it is certainly not appropriate in every case. For instance, it would be disastrous to suggest that a diabetic stop taking his or her insulin. Nor should it be suggested that individuals with diagnosed heart disease give up their prescribed heart medications.

In my own case, I found that the side effects associated with the medicines I had been prescribed were becoming as serious as the problems for which I was being treated. Considering my state of health, my age, and the opportunities for effecting a positive change, I knew the odds were in my favor: The fewer drugs I put into my body the greater my advantage would be in the long run. For years I had been aware of the fact that I had a tendency toward high blood lipid levels, but that fact had really meant little to me and the way I lived my life. When I was diagnosed with "hyperlipidemia" and "hypertension" with sodium retention, I felt some concern, but I never imagined that these medical terms would lead to a life sentence of dependency upon drugs. Hyperlipidemia is described as a genetic condition whereby the liver overproduces cholesterol, to the degree that the body is unable to excrete it properly. While the body actually requires a small amount of cholesterol for optimum functioning, it does not require much. It was further explained to me that because of my genetic predisposition, neither exercise nor dietary changes could effect enough of a change to make prescription drugs unnecessary. Because much of my life had been lived without any real health concerns, I did not accept very well the medical doctor's assumption that I was to rely exclusively on his professional opinion. While the stress that I felt regarding my present and future health was overwhelming, I didn't like the idea of giving so much control over my body to someone else.

My health crisis came to a sudden climax while I was experiencing radical changes in my life. Although the changes were constructive, they put strains on my energy—and to the body stress is stress. The endocrinologist I consulted did not provide me with a very positive picture for the future or any choices at all regarding my treatment plan. On the basis of my high blood pressure and cholesterol readings, I was given three prescription medications, one of which I was assured I would have to take for the rest of my life. Hypertension, hyperlipidemia, and edema: With all of these new disorders and medications, I felt overwhelmed. If ever I'd felt in my life that I had lost control, it was then. The snowballing effect of events in my life only added to the high stress level I had been experiencing. The medications were not covered by our group health care plan, and the cost climbed to nearly $300 per month. At that point, I wasn't complaining about the cost of my treatment plan—I would have been willing to pay whatever it took to be well again.

If I thought I felt a bit "down" with regard to taking all of these prescriptions, I still had more to learn about being down. Ten days or so into my treatment protocol (the daily doses of these three very potent drugs) new symptoms that I had never before experienced—shortness of breath and heart palpitations—began to occur. Serious suspicions regarding the drugs I was taking—Accupril, Bumex, and Mevacor—began to consume my thoughts, and I decided it was time to investigate the test findings associated with the clinical studies of them that had been made in accordance with FDA requirements. My stress level increased because of the new symptoms and my growing uneasiness regarding these drugs. My resolution to play an *active role* in my health care—including any decisions made about it—was an important shift for me. My determination to take back control over my body and actively participate in a treatment plan that felt right for me seemed instantly to release much of the stress I had been dealing with. It was a big responsibility, but one that I accepted willingly. Aware that many of the decisions I would be called upon to make would require a good deal of study and research on my part, I suddenly developed a new sense of meaning in my life. Life is all about challenge and opportunity, isn't it? As I undertook my new challenge, I felt certain that it was a mission I had been appointed to carry out.

As I began to study and learn about my conditions and prescriptions, I decided to keep my scheduled biweekly office visits with my endocrinologist. With each visit my resolve to reverse my health problems was strengthened. Slowly I incorporated healthier life-style practices into my life. As the doctor continued to insist that the drugs he had prescribed had "saved my life," my inner feeling of resistance grew stronger. What this doctor seemed to be expressing was a notion that my very being could not accept, that I should be content to accept life at the cost of health. The very essence of life itself for me was to be able to participate fully and contribute whatever I could. The very sense of anticipation for tomorrow, the energy, and the creativity that usually accompanied each new day were gone. The doctor's only interest seemed to be to promote acceptance and compliance in following his recommendations. Real communication and a sense of connection were nonexistent. My attempts to explore my health issues through questions always left me feeling frustrated and in conflict. What I didn't realize at the time was that my inner conflict was caused by the fact that I was not living in a manner that was compatible with my beliefs. My physician's philosophies regarding health and life were simply not consonant with my own, and I distrusted his method of treatment. When I asked him how long I would need to remain on the blood pressure lowering medication, he said: "Always, Eileen. You'll be taking these medications for the rest of your life." Prodding him a bit further, I found him unable to give me any convincing information that would allow me to accept his treatment method and prognosis; therefore,

I found no justifiable reason for remaining compliant to his particular brand of therapy. Before walking away, however, I insisted on a response to what I believed was a logical patient/doctor question. Looking him in the eye, I asked once again: "To what do you attribute this condition, doctor?"

His response was, "It's called essential hypertension."

"But what causes essential hypertension?" I persisted.

His answer was, "In essential hypertension, no identifiable cause can be found."

Rather than satisfying me, this diagnosis only strengthened my own belief, which was: If there was no apparent reason for my hypertension, there was no logical reason for having high blood pressure! The doctor's response to my next question really helped me to map out my course in life. Again I prodded, "Doctor, if there is no physiological cause for this problem which has become acute, why do you feel so strongly that it couldn't leave as it came, suddenly?"

A short pause preceded his reply. "You will always need to be on medication, Eileen."

"Why?"

His response—"Because"—was what finally sent me on my way.

Though his answer may have been acceptable to me as a child, I was no longer a child. While I had already begun my own pursuit of uncovering the truth (and the truth, it is said, will set you free) in order to find my way back to good health, I now stepped up my pace. The Serenity Prayer helped to keep me focused: "God grant me the serenity to accept the things I cannot change, the courage to change the things I can, and the wisdom to know the difference."

Believing that only I could change myself, I became deter-mined to accomplish just that. With determination and re-solve we can accomplish whatever we need to do. So it was that I became a Doctor of Naturopathy.

Today, I take many natural supplements, exercise as much as possible, avoid stress, eat healthy foods, and sur-round myself with positive friends. There are no prescription bottles in my medicine closet, and my blood pressure read-ings are normal. My sodium sensitivity problem is well man-aged, as is my cholesterol problem. My life, my health, and my career are now all integrated, and I live in harmony with the philosophies that are compatible with my beliefs.

Introduction

Today the average person is faced with a plethora of conflicting health information, some of it worrisome, some of it encouraging. Most of us are aware that an increasing variety of toxic synthetic substances are finding their way into our food supply and that the full consequences of this assault on our health cannot as yet be determined. We are also aware that alternative health options are becoming more widely available, that an active lifestyle, proper diet, and nutritional supplementation are known to have a positive impact on our health. For many of us, however, the proliferation of information—both on avoiding health risks and seeking alternative health options—has become a source of confusion and frustration.

This book is designed to make that information available in an easy-to-use, alphabetical format and deliver it to readers in concise, everyday English. It can be consulted as a quick reference for information on a specific additive, toxin, nutrient, or supplement, or as an overview of particular subjects, such as the role of diet and nutritional supplementation in maintaining health, or the possible consequences of consuming a tremendous number and variety of additives throughout our lives. Having convenient access to this kind of information makes it easier to become more conscious as consumers and to take greater responsibility for our own and our family's health.

Becoming an active participant in maintaining health has become an important issue for many reasons. As information becomes available about specific side effects associated with the toxins that assault our bodies on a daily basis, more and more people are seeking a better understanding of how to protect the human immune system from these hazardous chemicals. At the same time, health care costs have risen significantly in the past decade, while health care packages on average are too expensive for many people and provide for scarcely half of what they once covered. Even when treatment is available, many are learning firsthand that modern pharmaceutical medications and standard treatments are not necessarily a cure-all. The exorbitant cost of medical services, as well as the side effects and potential risks of medications, encourage the public to seek out alternative health solutions.

Despite conscious efforts to inform ourselves about healthier habits, most of us know very little about the chemicals contained in the food we eat every day. It may be tempting to believe that a kindly Big Brother is watching out for us, that law and government regulation adequately protect us, and that foreign substances would not be added to our food without our knowledge or consent. In fact, this is exactly what often happens. For example, chemicals that occur naturally in commonly eaten foods are frequently not required by the Food and Drug Administration (FDA) to be listed on the ingredient label, although these same substances can reach toxic levels when they are used as additives.

Chapter I lists and describes over a hundred of the most common additives found in food products, as well as chemicals present as residues in foods—including pesticides, hormones, and other synthetic substances that find their way into produce, meat, and other non-packaged foods. The information in this chapter will help familiarize readers with the characteristics of common additives so that they can make sense of food labels and determine which products are the safest to buy.

At the beginning of the first chapter you will find a description of the major types of additives. The separate alphabetical entries describe the characteristics of each food additive, its purposes, the types of products in which it is used, adverse reactions that have been associated with it, and usually, how the FDA has classified it or limited allowable residues in foods. Chapter I includes a separate listing of additives that are considered benign or beneficial.

Chemical compounds are added to many kinds of foods primarily to preserve them, enhance or maintain their customer appeal, or extend shelf life. It should be noted that in order to extend shelf life, many nutrients or vitamins must first be removed from the food. The manufacturer must then, as required by the FDA, "add back" one or more synthetic vitamins, usually at minimal levels. Some additives are used to prevent mold, fermentation, or bacterial or fungal growth. Others are used to preserve color, flavor, texture, moisture, and so on.

The FDA is charged with investigating all additives, pesticides, and so forth that make their way into food, but in many cases its information is limited. In weighing such evidence as it has, the FDA has conferred the status of "Generally Recognized As Safe" (GRAS) on many hundreds of additives and has set "tolerance" levels for pesticide and other residues, based on their use "at present levels." Many of these substances are proved or suspected carcinogens if consumed at higher concentrations. The cumulative and long-term effects of most additives and residues have not been determined, and the effect of consuming a daily "cocktail" of mixed chemicals has barely been considered. The FDA and a few consumer hotlines may be able to provide additional information on specific chemicals (*see* Consumer Resources, page 255).

Reactions to additives and residues depend upon the characteristics of the substance, the amount and frequency with which it is consumed, as well as individual sensitivity, age, state of health, or genetic weaknesses. Some of these chemicals have been linked to high blood pressure, vitamin deficiencies, kidney problems, and cancer. The very young, the ill, and the elderly are at higher risk for adverse reactions.

Of perhaps equal concern are the residues of the pesticides and fungicides that are sprayed onto produce crops, including much of the grain used for animal feed. The quantity of remaining pesticide residues are difficult to monitor throughout the food supply and can easily surpass those figures deemed "safe" for human consumption. As noted before, the FDA's tolerance levels for these chemicals do not take into consideration the combined toxic load of all the synthetic substances we consume from various other food sources. The accumulative effects of additives and pesticide residues can easily overburden the liver's ability to rid the body of toxins and thus weaken the immune system, especially among children, older adults, and others whose health is already compromised.

With good reason, many of today's shoppers are going the extra mile to pay more to purchase organic foods, which are raised without toxic pesticides, fungicides, or hormones—and further, in the case of processed foods, without dangerous additives. (*See* Consumer Resources for information on locating sources of organic foods.)

Chapter II discusses the essential nutrients required in the human diet for life and maintaining optimum health. Amino acids, essential fatty acids, minerals, and vitamins are covered in individual alphabetical entries, and a general explanation of each category of nutrients is provided. This chapter also covers digestive enzymes and explains how they act in the body. Chapter III gives information on natural food sources that provide these nutrients and gives guidelines for minimizing toxic residues when organic produce is unavailable or creates a strain on the budget. The chapter also offers substitutes for many of the "empty calorie" foods that have gained such prominence in the age of fast food. These foods are especially popular among children and are notably dangerous to our health because they fill us up without satisfying nutritional needs and are the foods that are most heavily loaded with additives. Chapter III also discusses food preparation and includes tips about maintaining safety as it concerns storing, preparing, and serving various foods.

Chapter IV is a guide to the nutrients and natural therapeutics that line the supplement sections of health food stores and that are found in increasing numbers in drugstores and supermarkets. These supplements are sold separately or in combination and in a form that is often more concentrated than can be found in natural food, herb, and mineral sources. The chapter on supplements discusses the importance of differences in individual needs for particular nutrients, as well as situations in which a particular supplement should be restricted or avoided entirely. The interaction of certain vitamins with various types of medications is discussed. It is stressed that supplementation should not become a form of self-medication resorted to without the guidance of a knowledgeable health care provider.

For centuries herbs have been the primary source of medicines throughout the world. Even in the present day, a large proportion of mainstream medicines are based on the active ingredients of medicinal plants, often manufactured synthetically. In recent years the general interest in herbs has revived, as researchers and alternative health providers have explored their uses. Today herbs are often regarded as a "do it yourself" method of maintaining health or treating illness. This approach can have very serious consequences. Chapter V covers most of the herbs that you will find in health food stores and specialty stores. It provides complete information on which herbs are non-toxic, their beneficial uses, and how and in what dosages they can be recommended. Cautions in the use of herbs are gone into in detail. Many herbs should be taken only with professional supervision by a naturopath or other practitioner knowledgeable in this field. Herbs have a very wide variety of effects and their use needs to be tailored to the individual. Some can be beneficial for one person, yet raise blood pressure or create a metabolic disturbance in another. Certain herbs taken in excess may cause liver disorders. Some particularly potent herbs may be toxic or even fatal and are not recommended under any circumstances.

In addition to providing information on specific substances—both helpful and harmful—it is hoped that this book will provide a general overview that will help readers look at health questions as a whole and make the kinds of choices across the board that support a healthy and sane lifestyle and diet.

Chapter I: Additives

ADDITIVES AND PROCESSING METHODS

Although additives now are employed for fifty separate uses, it is possible to group them loosely in the following categories:

ACIDS, ALKALIES, BUFFERS and NEUTRALIZERS ∽ Processed foods depend upon additives such as sodium aluminum phosphate or tartaric acid for maintaining the desired level of acidity or alkalinity. Colas depend largely upon phosphoric acid for their distinctive citrus taste. Other commonly used buffers and neutralizers include ammonium bicarbonate and potassium acid tartrate.

BLEACHING AGENTS ∽ Some bleaching agents, including benzoyl peroxide, nitrosyl chloride, and chlorine dioxide, are utilized for the bleaching of flours and as a flavor enhancing agent (benzoyl peroxide, however, only bleaches, without flavor enhancement.)

COLORING AGENTS ∽ These agents, both natural and synthetic, are employed as coloring agents because they provide consumer eye appeal that increases sales. On the downside, however, these coloring agents can hide the obvious inferiority of a particular product. When coal tar colorants were found in the 1950s to produce negative effects in laboratory animals (after having received the GRAS status in 1938), they were removed from the list of safe additives.

The federal government (in 1960) instituted a policy that required manufacturers to retest all artificial colors in order to determine the safety of continuing their use. Synthetic coloring agents include FD and C Blue No. 1, FD and C Citrus Red No. 2, and FD and C Red No. 40. Though these artificial coloring agents were shown to cause tumors in laboratory animals at the place of injection, the FDA does not find these data important, because these results were caused by injections, rather than by ingestion. Coloring agents that are "natural" include annatto, carotene, chlorophyll, saffron, and turmeric. These natural coloring agents (which have shown no negative testing results) can often be found in such foods as butter, cream, baked goods, candy, and soft drinks.

FLAVORINGS ∞ These additives comprise nearly two-thirds of all food additives used today. In the preparation of many foods (more often referred to as processing), flavor escapes. Some two-thirds (1,500) of these flavorings are synthetic, and only 500 are classified as "natural." They are incorporated in amounts ranging from a few to 300 parts per million. Extracts, essential oils, and oleoresins are taken from both natural and synthetic sources. They are commonly found in soft drinks, ice cream, bakery products, and other types of sweets.

HUMECTANTS ∞ These substances are utilized in certain confectionary products such as candy and shredded coconut for their ability to retain moisture and inhibit drying. Some preservatives belonging to the humectant category include sorbitol, glycerine, and propylene glycol.

IRRADIATION ∞ Radioactive isotope beams spray ionizing radiation on food that is passed by them on a conveyor belt. Radiation retards ripening and destroys certain bacteria and molds that cause spoilage. This process allows food to retain its attractive eye appeal and protects its taste for nearly two months. While irradiation does not make food radioactive, many other questions are left unanswered. For instance, can it destroy a food's nutrient content? Could there be any reason for concern regarding exposure of these radiolytic products that might lead to genetic problems? The FDA requires that foods which have been processed through irradiation indicate as much on the label; they must also display the flower in a circle, the international logo to indicate an irradiated product.

PRESERVATIVES and ANTI-SPOILANTS ∞ These additives help to prevent spoilage or are employed to lengthen the shelf life of a particular food. *Antioxidants* are preservatives used in fatty foods for the purpose of inhibiting distasteful flavors or odors. Antioxidants include butylated hydroxytoluene (BHT) and butylated hydroxyanisole (BHA), which are employed in breads, chips, crackers, and cereals. Other preservatives such as sodium propionate, calcium propionate, and sodium diacetate are utilized as *mold inhibitors*, and can be found in breads, cheeses, and pie fillings. *Fungicides* are preservatives that inhibit the growth of fungi. They are commonly used on citrus fruits. Another type of preservative used for maintaining the texture, flavor, and appearance of specific foods is classified under the category of *sequestering agents*. Sequestering agents such as EDTA, or

ethylenediaminetetraacetic acid, are utilized in the manufacture of soft drinks for their ability to inhibit toxic effects
from metals, and they are employed in dairy products to help
keep them fresh. Similar preservatives which provide several useful purposes in the manufacture of foods include
sodium, calcium, and potassium salts of citric, tartaric, and py-
rophosphoric acids; propyl gallate; sugar; salt; and vinegar.

PROCESSING AGENTS ∾ This category of additives includes
various subcategories, such as *sanitizing agents*, *chelators*, *clarifying agents*, *emulsifiers*, and *stabilizers*. Sanitizing agents are
employed for removing bacteria from a product, as are tannins, which are used to clarify wines. Maintaining smooth-
ness and a consistent texture is aided by the use of stabilizers and emulsifiers. Frequently used for such purposes
are monoglycerides, diglycerides, and lecithin. Calcium
chloride is an additive that provides more "body" to a product and is often found in canned foods. *Texturizers* often used
in the making of ice cream or frozen ices include agar-agar
and cellulose gum. Artificially sweetened drinks, including chocolate milk, require what is known as *texturizer gums*.
This type of texturizer prevents the cocoa particles in the
chocolate milk from falling to the bottom of the drink. This
desired cohesive texture is maintained through the use of
such additives as pectin or gelatins. Thickness is usually attributed to the product's sugar content. Sodium alginate is
yet another type of thickener.

VACUUM SEALING ∾ This is a method of preservation in
which partially cooked foods are sealed in vacuum containers (containing little or no oxygen) in order to prevent

spoilage. This type of food presently represents more than $400 billion annually. However, scientists have voiced concern regarding dangerous bacteria that could escape being killed during the "partial precooking," particularly the bacteria that cause botulism, which can remain alive in an oxygen-free atmosphere.

ADDITIVES

The following is a list of common food additives representing the types outlined above. Many of the following additives judged to be toxic in varying degrees.

ACETIC ACID ∽ This is an acid that occurs naturally in apples, cheese, cocoa, coffee, grapes, skimmed milk, oranges, peaches, pineapples, strawberries, and a variety of other fruits and plants. About 6 % acetic acid is contained in vinegar, while essence of vinegar is estimated to contain about 14 %. Because it is widely used as a flavoring agent in cheese and baked goods (including animal feeds), it is allowed in many "standardized" foods, but it is not required by the FDA to be listed on the ingredient label. A solvent for resins and gums and volatile oils, it can stop bleeding when applied to a skin cut. Acetic acid is also used in creams for bleaching freckles, as well as in hair dyes. Adverse reactions might include skin irritations such as hives, itching, and soreness, as well as an overabundance of organisms that might not respond positively to disinfectants. In a potent form (without much water), it becomes extremely corrosive, with fumes capable of causing a lung obstruction.

At potencies of less than 5%, acetic acid in a diluted liquid is somewhat irritating to the skin and causes cancer in laboratory animals when given either by mouth or through injection. This information was provided to the FDA in a document several years ago.

ACETYLATED ∾ "Acetylated" refers to a process in which an organic substance is heated with *acetic anhydride* or *acetyl chloride* in order to remove its water. An acetylated coating is used on candy and other foods for the purpose of retaining the product's moisture. The fumes of acetic anhydride cause irritation and death of tissues; products containing this chemical must carry a warning on the label against contact with the skin or eyes.

ACHILLEIC ACID ∾ See **Aconitic acid.**

ACONITIC ACID ∾ This acid is also known as *citridic acid, equisetic acid*, and *achilleic acid*. Used as a flavoring agent, it is found naturally in beetroot and cane sugar. Commercial aconitic acids are generally manufactured through sulfuric-acid dehydration of citric acid. Aconitic acid is a component of fruit, brandy, and rum flavorings. It is often added to such foods as ices, ice cream, beverages, candy, liquors and other food products. It is also used in the manufacture of plastics.

AGROSOL ∾ See **Captan.**

ALANINE (B-, L-, and DL-) ∾ Obtained from a protein, the colorless crystals of alanine are thought to be a nonessential amino acid. Alanine is a substance that has been used in microbiological studies and as a food supplement in its L and DL forms. It has received the GRAS status from the FDA as a food additive. However, studies have linked it to cancer of the skin and other tumors in mice. The FDA data bank, Priority Based Assessment of Food Additives (PAFA), has toxicology information regarding this chemical.

ALGINIC ACID ∽ An odorless, tasteless, gel-like substance, alginic acid is extracted from seaweed. It is commonly used as a stabilizer in ice cream and other dairy products, as well as in beverages, icings, and some salad dressings. Alginic acid is utilized as a defoaming agent in processed foods, and it absorbs many times its weight of water and salts. It is also used in the paper, textile, and cosmetics industries. Similar to albumin or gelatin, alginic acid dissolves in water to form a heavier liquid. According to the World Health Organization (WHO) Committee on Food Additives, in a three-month study of rats, 15% alginate in their food produced abnormalities in the lower intestines and bladder. Other specific findings included calcium deposits in the kidney and a slight inhibition of growth. Larger doses induced a laxative effect. The committee did not suggest an Acceptable Daily Intake (ADI); thus alginic acid is allowed as an additive in certain foods.

∽ ∽ ∽

ALLYL SULFIDE ∾ As a synthetic flavoring, it is often added to ice creams, soft drinks, candy, baked goods, and meats; it is found naturally in garlic and horseradish. It has produced negative reactions in the respiratory tract and eyes and can easily be absorbed through the skin. Excess exposure can produce loss of consciousness, and continued exposure has been associated with liver and kidney disorders. The FDA's data bank, PAFA, has more information for the interested consumer regarding this additive.

ALUM ∾ See ***Aluminum sulfate.***

ALUMINUM ∾ A solid white crystalline metallic element, aluminum has often been used as an additive in foods and cosmetics. The ingestion or inhaling of aluminum can have severe negative effects on those individuals who suffer from kidney or lung diseases. Aluminum has been found in the brains of Alzheimer's victims, and a suspected link to the disease has been suggested, although more study is indicated.

ALUMINUM SULFATE ∾ Commercially sold under the name alum, this chemical substance is colorless, water soluble, and sweet tasting. These crystals have a drying effect in the mouth. Aluminum sulfate has long been utilized in the preparation of pickles and in the packaging of other foods, such as potatoes and shrimp. It is used as an antiseptic, in antiperspirants, and as an astringent. In an antiperspirant it may cause irritation or allergic reactions. Said to affect reproduction, it is mildly toxic by ingestion or injection. In 1980, the report to the FDA of the Select Committee on

GRAS retained aluminum sulfate on the GRAS list. The only restrictions were that it should be used under "good manufacturing practices."

ARSENIC ∽ Crystals of arsenic are very dark to nearly black in color. Arsenic has been used in animals to promote growth. The FDA sets the level of tolerance at 0.5 ppm in muscle and 2 ppm in raw edible by-products. In eggs from chickens and turkeys, the FDA sets the limitation at 0.5 ppm, with a tolerance level of 2 ppm in liver. Arsenic, however, is a human carcinogen, poisonous to all bodily organs. Ingestion can cause toxic effects to the skin and in the intestinal tract, as well as birth defects.

ASPARTAME (NutraSweet®) ∽ A chemical compound 200 times sweeter than sugar, it occurs naturally in *aspartic acid* and the amino acid *phenylalanine*. This additive has been the subject of much controversy regarding its safety. The G.D. Searle Company first requested FDA approval in 1973. However, there was continuing concern as to whether aspartame could cause brain damage. An independent audit of aspartame was set up by the FDA to study its potential effects for two years. The FDA appointed an "Expert Board of Inquiry," which reviewed the independent team's results and concluded that the evidence uncovered did not support the conclusion that aspartame might kill brain cells or cause other nerve damage. However, since individuals diagnosed with phenylketonuria (PKU) are advised to avoid foods containing phenylalanine (one of two components contained in aspartame), NutraSweet® should be avoided as well. The Board of Inquiry recommended approval of aspartame be withheld pending long-term animal testing to rule out aspartame as a cause of brain tumors. The FDA's Bureau of Foods concluded that the board's cautions were unfounded, and in October of 1981, aspartame was approved for use as a tabletop sweetener and for use in certain dry foods. Since that time, negative information has continued to surface from various reputable sources. The Arizona Department of Health Services announced in 1984 that they were testing soft drink beverages that contained aspartame to see whether it deteriorated into toxic levels of methyl alcohol under storage conditions. The Arizona Health Department was responding to Arizona State University's Food Sciences Research Laboratory, whose director submitted a document alleging that greater-than-normal temperatures could lead to a serious breakdown in chemical composition. But the FDA representatives contended that ordinary fruit juices contain high levels of methyl alcohol and that it believes concerns regarding decomposition products are unfounded. Aspartame lowers the acidity of urine; so it is said, theoretically, to weaken the urinary tract, making it vulnerable to infections. David Steinman (an expert in the field of food and water safety) has written, "Aspartame is a synthetic compound consisting of two amino acids—aspartate and phenylalanine— that are held in chemical bond by methanol, also known as wood alcohol. Aspartame is a very tiny molecule that the body cannot metabolize slowly; it goes directly into the bloodstream. When the aspartame molecule enters the bloodstream, the chemicals in aspartame break down into their constituents. Methanol is poisonous when found alone, as it is after aspartame enters the bloodstream and breaks down. Although we eat many foods that contain methanol, it usually is found with ethyl alcohol, which neutralizes the methanol. In aspartame the ethyl alcohol is absent" (Steinman et al.). In 1988 the Mexican government prohibited soda and food processors from using "nutra" in the brand name because it felt it was "misleading." They insisted on the following label for aspartame: "This product should not be used by individuals who are allergic to phenylalanine. Consumption by pregnant women and children under 7 years is not recommended. Users should follow a balanced diet. Consumption by diabetics must be authorized by a physician" (quoted in Winter 1994, p. 71).

AZODICARBONAMIDE ∾ Used in flour as a bleaching agent, azodicarbonamide ranges in color from yellow to red and is an insoluble crystalline. At last notification (1993), this additive was still under review for its possible long-term effects; however, it is allowed as a food additive.

BENTONITE ∾ A colloidal clay (aluminum silicate), bentonite has the ability to absorb huge amounts of water. Its many uses as an additive include as a thickening agent and a colorant for wine. If given intravenously it can cause blood clots and possibly tumors. The FDA's data bank can provide more toxicology information on this additive.

BENZOIC ACID ∾ A naturally occurring compound found in cherry bark, tea, anise, and raspberries, benzoic acid is used for its antifungal properties and as a preservative. In the early 1600s it was discovered and described as a substance in gum benzoin. It is used as a food preservative for chocolate, fruit and nuts, and in flavorings for desserts, candy, beverages, and chewing gum. It is allowed as a food supplement at 0.1%. This chemical compound can be somewhat irritating to the skin and may trigger allergic reactions in susceptible individuals, such as asthmatics or those sensitive to aspirin. Given the GRAS status by the FDA Select Committee in 1980, benzoic acid has no limitations governing its use other than to use "good manufacturing practices."

BHA ∞ *See* **Butylated hydroxyanisole.**

BHT ∞ *See* **Butylated hydroxytoluene.**

BIFENTHRIN ∞ A synthetic pyrethroid (a compound related to oily liquid esters that occur in chrysanthemums) bifenthrin has been used for controlling insects and mites. The FDA sets tolerances for this pesticide at 0.02 ppm in milk, 0.10 ppm in fat, meat, and meat by-products, and 0.50 ppm in cottonseed. This pesticide is considered to be less toxic than many other types of pesticides used today.

BIS(TRIS[BETA, BETA-DIMETHYLPHENETHYL]TIN)OXIDE ∞ This chemical is used as an insecticide in animal feed, as well as on dried fruit and in fruit products, including citrus pulp. The FDA residue tolerance is 20 ppm for prunes and raisins. For dried apple pomace (pulp), the tolerance level is 75 ppm, and 35 ppm for dried citrus pulp. Also known as "Bendix," this insecticide is considered mildly to moderately toxic either by ingestion or through contact with the skin. Allowances are three to five times higher when used in certain types of animal feed.

BOVINE SOMATOTROPIN (BST) ∽ First brought to public attention in 1994, this synthetic growth hormone is used to stimulate a higher production of milk in cows. The FDA, the manufacturers of the hormone, and some experts believe that this hormone is one that naturally occurs in cow's milk; therefore, supplementing cows with more of this hormone should have little effect. Small farmers, some scientists, and many consumer groups, however, oppose the use of this hormone. Their objections are numerous. BST aggravates inflammation of the udder in cows and may have unknown negative effects on humans. Also, sufficient reason for its use has not been adequately explained; because milk production surpluses have already affected government pricing, the idea of increasing milk production does not seem economically justifiable.

BROMINATED VEGETABLE OIL ∽ Bromine is a heavy, volatile, corrosive liquid element that is used for emulsifying oils. Its color varies from pale yellow to dark brown, and it has a "fruity" smell. Brominated vegetable oil is mixed with other oils of low density to produce a higher density oil that is easily emulsified. It is often an ingredient in flavored beverages, including sodas and ice cream. Brominated oils were placed on the FDA's "Suspect List" several years ago, so that the search and review of this additive's toxic potential would continue. It is still in use as an additive, as of this writing.

∽ ∽ ∽

BST ∞ *See* **Bovine somatotropin.**

BUQUINOLATE ∞ An anti-parasitic drug, buquinolate is almost exclusively used in chicken feed. Limitations for residues of this chemical have been set by the FDA at 0.4 ppm in uncooked liver and chicken skins. Limitations are set at 0.5 ppm in egg yolks and 0.2 ppm in whole eggs.

BUTYLATED HYDROXYANISOLE (BHA) ∞ This chemical substance is used in various foods as an antioxidant, preservative, emulsifier, and stabilizer. It can be found in chewing gum, cereals, baked goods, soup bases, candy, ice cream, and dried potato flakes. In enriched rice, animal fats, and shortenings, it functions as an emulsion stabilizer. Almost solid in consistency, BHA is insoluble in water. Depending upon how it is utilized, different limitations are applied. For instance, in fats and oils, only 0.02% is allowed, but in emulsion stabilizers, up to 200 ppm is the tolerance level set by the FDA. Though this additive can cause allergic reactions and has been linked to abnormal animal behavior in some studies, some experts believe that it is less toxic to animal kidneys than the related chemical known as BHT, or butylated hydroxytoluene.

BUTYLATED HYDROXYTOLUENE (BHT) ∞ This substance, used as an antioxidant, preservative, emulsifier, and stabilizer, is similar to butylated hydroxyanisole (BHA), but possibly more toxic. The American Societies for Experimental Biology, which advises the FDA on food additives, has suggested that more long-term studies be carried out on BHT (at presently allowable limitations), in order to determine how it will interact with the ingestion of steroid hormones and oral contraceptives. There has been some concern regarding whether BHT could transform other substances into toxic, cancer-causing agents. BHT is banned as a food additive in England. BHA and BHT are commonly used in the United States as a preservative in many widely sold foods, including cereals and shortenings. Laboratory studies in mice (using small levels of BHA or BHT in their food) produced offspring that often displayed abnormal behavior.

BLUE NO. 1 (FD and C) ∞ A coal tar derivative, Blue No. 1 is used as a food dye or coloring in processed snack foods, beverages, and candy. While it may not be listed on your ingredient label, it is a dangerous compound and a potential carcinogen. This food coloring, banned in France and Finland, may cause chromosomal damage.

BLUE NO. 2 (FD and C) ∞ Similar in chemical structure to Blue No. 1, Blue No. 2 is also a coal tar derivative. Also found in many "junk foods," it is believed by many that Blue No. 2 may cause brain tumors in laboratory animals. Sadly, this chemical compound is often found in pet foods. It is banned in Norway.

CALCIUM IODATE ∞ A white powder with little odor, it is used as a dough conditioner and as an oxidizing agent in baked goods. Capable of causing allergic reactions, calcium iodate is a nutritional source of iodine. The FDA toxicology report regarding this additive can be obtained by calling their consumer hot line.

CALCIUM METASILICATE ∽ This additive is insoluble in water. Limitations for tolerance levels have been set by the FDA at 5% in baking powder, and 2% for table salt. Calcium metasilicate has been utilized as an ingredient in antacids, a filler for paper coatings, and an additive for various foods. The residue dust from this chemical causes irritation upon contact.

CALCIUM OXIDE ∽ Used as a dough conditioner in baked goods and a neutralizing agent in dairy products, calcium oxide is also utilized industrially in such products as bricks, stucco, and insecticides. A strong irritant, this additive can seriously damage the skin or mucous membranes. The Select Committee on GRAS substances declared in 1980 that this substance should keep its GRAS status, with no limitations other than the following of "good manufacturing" practices.

CAPTAN ∽ This is a fungicide that is also manufactured under the names Merpan, Agrosol, Orthocide, and Vanicide. It is isolated from tetrahydrophthalmide and trichloromethylmercaptan. Almost completely insoluble in water, it is used for preserving fruit and on almonds as a fungicide. The residue tolerance allowed by the FDA is 50 ppm in raisins and 100 ppm on corn seed for cows and pigs. It is also used on vegetables such as broccoli, beans, carrots, potatoes, on fruits such as peaches and strawberries, as well as on many other foods. Compared to other fungicides in use today, captan might be considered less toxic than many. However, it can induce loose bowels and weight loss and is an irritant to the lungs. Some suspect that it can cause human birth defects and cancer. It is recommended that this fungicide be avoided by pregnant women.

CARMINE ∾ This is a red pigment derived from the cochineal insect. Commercially available in the form of an extract, cochineal extract is not allowed in its undiluted, potent form. Carmine, however, was at one time an ingredient in bakery products, applesauce, and spices. It was banned by the FDA when an outbreak of salmonellosis occurred; one death and the serious illness of 20 other patients in a New England town were traced back to carmine. It was an ingredient in a hospital's diagnostic solution, used for testing organs. It was also available in various commercially marketed cosmetic products. The FDA has not yet done a toxicology investigation on all of the available data relative to carmine.

CARVACROL ∾ This additive has an active component that occurs naturally in many oils, such as those of oregano or marjoram. Utilized as a synthetic fruit, mint, or spice flavoring, it can be found in candy, soft drinks, and ice cream. A highly toxic corrosive, carvacrol taken by mouth can induce death. One gram can trigger respiratory episodes and heart failure. PAFA, the FDA's data bank, can provide more toxicology information regarding this toxic substance upon request.

CEPHARPIRIN ∾ Limitations on residue amounts for this chemical substance have been determined at 0.02 ppm in milk and 0.1 ppm in uncooked edible tissues of dairy cattle. Cepharpirin is a drug that is approved for use in animals by the FDA. This is an example of how we ingest "indirect" toxins, as it can be found in beef cattle and in dairy cows. Elevated temperatures can cause this substance to emit toxic fumes.

CHLORAMPHENICOL ∽ Potentially lethal when given by injection, chloramphenicol is moderately toxic when taken orally. Associated with specific types of anemia and leukemia, chloramphenicol is known to be a human carcinogen. It is used as an antibiotic for pork, beef, and lamb. A story in the *New York Times* (October 24, 1993) mentioned that the Environmental Protection Agency (EPA) was conducting a study into the possibility of this drug's connection to cancer.

CHOLINE HYDROCHLORIDE ∽ Transparent to white in color, this chemical component is used in animal feed as a mold inhibitor. Moderately toxic to humans through ingestion, it could be considered a carcinogen. The Office of Consumer Affairs has placed this chemical on the "Consumer's Right-To-Know List."

CITRIC ACID ∽ *See* **Aconitic acid**

COCHINEAL ∽ *See* **Carmine.**

COTTONSEED OIL ∽ In tests this oil has been shown to contain as many as 10 different pesticide residues, including DDT and dieldrin. Commercially it has been used for such products as bedding, cosmetics, salad dressings, and packaged foods. It is high in saturated fats and capable of causing allergic reactions. It is used by many restaurants and can be found in such prepared foods as chocolates, candies, and many fried foods, such as doughnuts or potato chips.

CYCLAMATES ∽ Marketed in two forms, *sodium cyclamate* and *calcium cyclamate*, this additive was utilized mainly as a sweetener to replace refined white sugar and was found in everything from chewing gum to soft drinks. More than 175 million Americans were ingesting cyclamates in substantial amounts. In 1969 when it was discovered that cyclamates had caused bladder cancer in laboratory rats, they were taken off the market.

DEHYDROACETIC ACID (DHA) ∞ Sodium dehydroacetate is a powdery compound used as a preservative in several prepared vegetables. Due to its versatility, it is also employed as an antibacterial and antifungal agent, and can be found as an ingredient in bathroom products such as toothpaste and shampoo. This chemical agent is capable of causing serious damage to the kidneys and other organs because of its ability to behave as a blocking agent. For that reason, it is prohibited from edibles. More information regarding this additive can be obtained from the FDA.

DELTAMETHRIN ∞ This pesticide, which includes cyanide, has been used in tomato products. Tolerance levels determined by the FDA have been set at 1 ppm.

DHA ∞ See *Dehydroacetic acid.*

DIACETYL ∞ Occurring naturally in many foods, including cocoa and coffee, diacetyl is used as a flavoring for baked goods, margarine, ice cream, and other artificially flavored products. This flavoring compound is commercially manufactured through a fermenting process and used widely in everything from fruits to liquors. Diacetyl has been approved by the Department of Agriculture for flavoring margarine "in amounts designed for that purpose." The FDA assigned

GRAS status to diacetyl some years ago. However, this potent compound has been directly associated with cancer when given to laboratory animals in experimental studies. More information is available from the FDA upon request.

DIALIFOR (Torak) ∞ Dialifor is yet another insecticide that is added to animal feed grain. Tolerance levels determined by the FDA are up to 2 ppm in raisins, 40 ppm in dried apple pulp, and as much as 115 ppm in dried citrus pulp. This chemical is highly poisonous by ingestion or through skin contact.

DISTARCH PROPANOL ∞ A modified starch, this additive was given GRAS status by the FDA before its review was completed. Although there is no available evidence to validate concerns regarding the use of this compound at current levels, uncertainties exist that require additional studies.

DYES ∞ Among the many food dyes tested and believed to be "permanently safe" are those manufactured from coal tar derivatives; thus, they are suspected carcinogens. Since 1906, when the first comprehensive legislation for food colors was put into effect, hundreds of food colors have been tested and listed as safe or banned. Many have been placed on a temporary list as "provisionally safe" and allowed, while only a few have made it to the status of permanently approved for addition to food products. Even many of those on the permanent list, such as FD and C Blue No. 1, are associated with allergies and malignancies in laboratory animals. Controversy over the safety of food colors continues. Organizations such as Ralph Nader's Public Citizens Group argue that *any* level of cancer risk is unacceptable.

ENDOSULFAN (Thiodan) ∾ Similar in its chemical structure to the pesticide DDT, endosulfan is manufactured from a combination of methane and benzene. Used in the growing of tea and vegetables such as carrots, spinach, and tomatoes, the FDA limits its use to 24 ppm on dried tea. A substance considered to be toxic to humans, endosulfan can be absorbed through the skin or by inhalation. Endosulfan is one of many compounds in the environment that is said to imitate the human hormone estrogen, and high levels of estrogen have increasingly been proven to increase the risk of some types of cancer, including breast cancer.

EQUISETIC ACID ∾ *See* **Aconitic acid.**

ERGAMISOL ∾ *See* **Levamisole.**

GLUCONIC ACID ∾ *See* **Sodium gluconate.**

HYDROGENATED OILS ∾ Hydrogenation is a process whereby hydrogen atoms are added to a liquid vegetable oil and placed under extreme pressures, which turns these vegetable "liquids" into semisolid saturated fats. Most people know by now that a small amount of unsaturated fats in our diet is beneficial. It is saturated fats that can eventually cause harm to the body. Commercially, these oils benefit only the manufacturer and are opted for because of their ability to lengthen the shelf life of many of their products. Hydrogenated or partially hydrogenated oils are often one of the ingredients listed in such foods as doughnuts, cakes, pies, and various other processed foods. Many researchers now hold these trans-fatty acids responsible for altering body cells, making them vulnerable to free radicals. Hydrogenated fats have been suspected of being involved in several types of cancer. It is wise to avoid them!

KABAT ∽ An insect inhibitor, kabat is used in the feed of barnyard animals, as well as in many processed foods that are sold for human consumption, including rice, rye, wheat and oat cereals, and many dried fruits. The FDA limits residue levels for this insecticide (also known as methoprene) at 10 ppm for the foods just mentioned. Kabat can cause skin irritations and is considered to have a low-level toxic effect when ingested. In experimental tests of kabat upon laboratory animals, cell mutations have been noted.

LEVAMISOLE ∽ This is a compound utilized in warding off parasites in beef and pork. It appears to strengthen a compromised immune system in humans. In the early 1990s, there was serious concern regarding its cost. Although veterinarians were paying $14 for the drug, cancer patients were being charged as much as $1,500. The manufacturer of the drug argued that the cost of the drug to humans was determined by the need to factor in the cost of hundreds of ongoing studies involving some 50,000 patients. Limitations for residues were set by the FDA at 0.1 ppm in cattle and pigs. Some of the effects upon humans when taken into the digestive system have included skin eruptions, elevated temperatures, and coma.

MELAMINE ∽ This compound is employed in the production of paper and paper products, including food-packaging materials. It has been linked to skin problems and is considered to be a mucous membrane antagonist. In some studies, this toxic substance was shown to be a carcinogen.

MENTHONE ∽ A flavoring agent found in raspberries and peppermint oil, menthone is manufactured as an artificial flavoring commonly used in such products as ice cream, pies, cakes, doughnuts, chewing gum, and soft drinks. It has been linked to gastric distress.

MERPAN ∽ See *Captan.*

METHANE DICHLORIDE ∽ This chemical compound, which has an odor similar to that of chloroform, has several uses in the food business, including extracting chemicals, diluting food colors, and decaffeinating coffee and other products. The FDA allowance for residues of this product is set at 30 ppm for spices, about 2 ppm in hops extract, and 10 ppm in coffee. Inhalation or introduction of this agent into the digestive tract of humans has been shown in experiments to cause cancer. Other human reactions to this drug include interrupted sleep patterns, numbness, and heart rate abnormalities.

MONOGLYCERIDES and DIGLYCERIDES ∽ Although these glycerine derivatives occur in nature, the particular types of mono- and diglycerides that are used as food additives are usually synthetic. They are employed as emulsifiers and defoaming agents and are found in a wide variety of food products, including baked goods, milk products, margarine, shortening, candy, and beverages. The results of one laboratory study involving a specific type of monoglyceride produced stunted growth in laboratory mice. Other findings show that this agent lowers the body's ability to metabolize essential fatty acids (EFAs). (*See* Chapter II for information on EFAs.) Studies involving similar glycerides produced reproductive abnormalities in laboratory rats. Despite the negative results of many such studies, these chemical agents are still in use. Caution regarding frequent consumption of foods containing these chemicals is strongly recommended.

MONOSODIUM GLUTAMATE (MSG) ∞ This chemical is marketed as an accent for enhancing the flavor and zest of certain dishes, and it is also an additive in many prepared and processed foods, to intensify flavor in meats, condiments, soups, candy, and many other foods. It is often associated with "Chinese restaurant syndrome"; reactions of sensitive individuals include chest pain, numbness, and headache. Laboratory studies of animals showed changes such as irritability, depression, and reproductive dysfunction. Baby food companies voluntarily eliminated MSG from baby foods, but its GRAS status remains.

MSG ∞ See **Monosodium glutamate.**

NICKEL SULFATE ∞ A crystalline compound often included in the manufacture of cosmetics and hair dyes, this nickel salt is also considered a mineral supplement (up to one milligram per day). In nature, it is a striking green color. When ingested it has induced vomiting and is poisonous in large quantities. Other toxic effects include blood vessel weakness, depression, and brain and kidney damage.

NITRATES and NITRITES ∽ Compounds containing both types of substances are used in cured meats for maintaining a vibrant red color. When combined with the stomach's own natural chemicals ("amines"), these additives produce potent carcinogens known as "nitrosamines." They are commonly used in the processing of such foods as bacon, sausages, hot dogs, luncheon meats, and many other foods. It is suggested that when these types of foods are eaten, a large dose of vitamin C be taken, because this vitamin can assist the body in flushing out these toxic chemicals.

NUTRASWEET® ∽ *See* **Aspartame.**

OCTADECENOIC ACID ∽ *See* **Stearic acid.**

OLESTRA® ∽ Better known as *sucrose polyester*, this compound is a non-digestible plastic that was created to replace the fat in such food favorites as french fries, potato chips, doughnuts, and other snack foods. In its favor is the fact that it contains no calories, but it has been reported in laboratory studies to cause liver dysfunction in animals. Warning labels on Olestra®-containing products warn consumers that this compound can cause digestive disturbances such as abdominal cramping and loose stools, and labels caution that it can inhibit the absorption of vitamins and other nutrients.

ORTHOCIDE ∽ *See* **Captan.**

PARABENS ∽ Methyl parabens and propyl parabens are regularly incorporated into many foods for their preserving effect and as mold inhibitors. These two compounds, which are similar in chemical structure, do not occur in nature. They are said to be safe when limitations of 0.1% restrictions are followed. They are often used in frozen dairy products, processed fish food products, oils, and bottled sodas. These compounds are used at levels of up to 1,000 ppm in other types of foods, including tomato puree, pickles, and other processed foods. Laboratory animals who were given these chemicals bore offspring with birth abnormalities.

PHENOL ∽ Originating from coal tar, phenol is used as an ingredient in producing various other additives and food-processing compounds. Coal tar is obtained from bituminous coal and is composed of many poisonous chemicals, including naphthalene, pyridine, and benzene. Coal tar derivatives have caused cancer in animals. Even in minute quantities, phenol in the digestive system can cause serious or fatal results. Nausea, paralysis, convulsions (or coma), and heart failure are just a few of the acute problems brought on by phenol.

PHENYLACETALDEHYDE DIMETHYL ACETAL ∽ An aromatic synthetic fruit, spice, and honey flavoring, this chemical is commonly used to flavor chewing gum, soft drinks, ice cream, candy, and baked goods. In the digestive system, this compound is moderately toxic.

PHOSALONE ∽ This is an insecticide used to eliminate insects and mites. The FDA has set the tolerance limits for its residue from 8 ppm to 85 ppm for dehydrated fruits, with the highest allowance for dried apple pomace (pulp).

PHOSPHOROUS OXYCHLORIDE (phosphoryl chloride) ∞ This chemical is used as a bleaching agent, solvent, and starch modifier. Since it may be used in the preparation of other compounds, it may be present in an end product without appearing on the label. Inhaling the fumes of this chemical can cause pulmonary edema.

PINE TAR OIL ∞ Synthetic flavoring isolated from the pine tree, pine tar oil is used for dissolving other compounds and as a disinfectant. It is used in licorice flavorings for many types of "junk" foods, such as ice cream and candy. As an oil, it is used in some bath salts. In concentrated form, pine tar oil is known to induce dizziness and nausea when taken into the digestive system and to cause skin irritation and allergic reactions.

POLYACRYLAMIDE ∞ Solid in consistency and soluble in water, this polymer is employed in modified form in the process of clarifying cane sugar, in manufacturing soft gelatin capsules, and in washing fruits and vegetables. This substance is also utilized in the manufacture of nail polishes and plastics. Polyacrylamide in its pure form is highly toxic. It is absorbable through the skin and can cause paralysis of the nervous system.

POLYETHYLENE ∞ This lightweight thermoplastic is produced from petroleum gas, or the dehydration of alcohol. Polyethylene is part of a chemical family known as polymers. A good deal of this chemical's value derives from its resistance to other chemicals, low absorption of moisture, and insulating properties. This chemical is used in packaging, cattle feed, and as an ingredient in chewing gum. In low concentrations, it is said to be relatively safe. However, in laboratory animals that were given large amounts of polyethylene, cancers developed.

POLYGLYCEROL ESTER ∾ This additive is one among many derived from saturated and unsaturated fats. It is an ester produced from edible fats, fatty acids, and oils, hydrogenated or non-hydrogenated. It is extracted from lard, tallow, palm, corn, peanuts, and soybeans. These types of esters are used in plasticizers, cosmetics, and as food emulsifiers. The FDA data bank (PAFA) can provide more information to concerned consumers.

POLYOXETHYLENE (20) SORBITAN MONOSTEARATE ∾ This emulsifier and flavor dispersing agent is also known as *polysorbate* (60). It is commonly used in prepared and processed foods such as cake mixes, chicken bases, gelatin desserts, and a variety of other foods. It is also used in some vitamin supplements, beverage mixes, and in substitutes for milk and cream. As of this writing, information regarding limitations for using this chemical is not available. Several years ago, the FDA requested that more studies be made regarding this food additive.

POLYSORBATE (60) ∾ *See **Polyoxethylene (20) sorbitan monostearate.***

POLYVINYL ALCOHOL ∾ A synthetic resin used to modify the color of eggshells, this polymer is derived from polyvinyl acetates through the process of replacing the acetate groups with hydroxyl groups. The FDA allows no penetration of polyvinyl alcohol through the eggshell. The International Agency for Research on Cancer (IARC) found this chemical to be a carcinogen. However, though this substance is "allowed" as a food additive, it is not presently known to be used for that purpose.

POLYVINYLPYRROLIDONE (PVP) ∽ This chemical compound is used in some dietary products and as a clarifying agent in vinegar, wines, and beer. The limitations for residues of this compound have been set by the FDA at 6 pounds per 1,000 gallons in wine and 10 ppm as a clarifying agent in beer. It is used as an additive in non-nutritive sweeteners, with tolerance levels of 60 ppm. Other products that might include this compound are vitamin and mineral concentrates. Ingestion can cause impairment to the kidneys and lungs. The available data regarding this chemical suggest that there is sufficient reason to suspect that residues might remain for anywhere from months to years in the lungs of people who have been exposed to it. Additionally, further exposure from non-food products, such as the PVP contained in hair sprays, increases the potential risk of impairment.

POTASSIUM SULFATE ∽ A whitish powdery substance, potassium sulfate is found in nature only in combination with sodium sulfate. It was developed for use as a food flavoring and as a salt substitute. It is also used in the brewing industry and as a fertilizer. In 1980 the Select Committee on GRAS reported to the FDA that this chemical substance should continue to remain on the GRAS list, without any limitations. Large doses have caused gastrointestinal bleeding.

PROPASIN ∾ *See* **Propazine.**

PROPAZINE ∾ Sold as *Prozinex* and *Propasin*, this compound is used on animal food crops for inhibiting overgrowth of weeds. In laboratory experiments, this chemical has caused tumors.

PROZINEX ∾ *See* **Propazine.**

PSEUDOPINENE ∾ A synthetic flavoring agent used in various types of food, pseudopinene is considered a low-level toxin if introduced into the digestive system.

PVP ∾ *See* **Polyvinylpyrrolidone.**

SODIUM DIHYDROGEN PHOSPHATE ∾ This chemical is used as an emulsifier, buffer, and dietary supplement. It serves as a poultry wash and as an additive to cheeses, meat and poultry products, and soft drinks. The FDA has set tolerance limitations for this product in meat products at 5%. It is toxic when ingested.

SODIUM GLUCONATE ∾ Also known as *sodium salt* or *gluconic acid*, this food additive is used as a preservative to prevent physical and chemical changes in food products and beverages. Not much more is known at this time regarding this chemical, other than that it was granted GRAS status in 1980.

SODIUM MONO- AND DIMETHYL NAPHTHALENE SULFONATE ∾ Employed as an ingredient in cured fish and meats, this chemical is an anti-caking and lye peeling agent. The FDA's Consumer Hotline may have more information regarding this compound.

SODIUM PROPIONATE ∞ This colorless crystalline compound is used as a preservative for inhibiting the growth of mold and fungi in baked goods, gelatin, frostings, and confections. It has been recommended for the treatment of fungal skin infections, but it can often trigger allergic reactions. There is not much information available regarding the use of this chemical at this time.

SODIUM SALT ∞ See *Sodium gluconate.*

SODIUM SILICATE ∞ This chemical is used as an anti-caking agent in soaps and creams and as a preservative for eggs. It can be irritating to the skin and mucous membranes. If taken by mouth, it will induce diarrhea, nausea, and vomiting. This chemical has received GRAS status from the FDA, with no limitations other than to use good manufacturing practices.

SODIUM SULFITE ∞ This light-colored, powdery additive is described as having a sulfur-like taste. It is utilized as an antiseptic, antioxidant, and preservative. In beverages, it works as a bacterial inhibitor and blocks fermentation in sugars and syrups. It is a commonly used antioxidant in processed foods subject to browning, such as fruits and peeled potatoes. Non-food uses include everything from hair dyes to the bleaching of silk and wool, and the developing of film. Foods and beverages containing sulfites may release sulfur dioxide gas, which can trigger asthmatic attacks. Sulfites have been linked to such problems as nausea, stomach irritation, skin rashes, and edema in sensitive individuals. Individuals with impaired kidneys or livers may be unable to produce the enzymes necessary to break down this chemical in the body. Sulfites have the ability to destroy thiamine and therefore are not added to foods that are sources of the B vitamin complex. The Select Committee on GRAS stated in their final report to the FDA that this substance is safe when consumed at present levels. They stated further, however that if consumption levels increase dramatically, additional information will be necessary to determine tolerance and limitation levels. Other sulfites commonly used as preservatives are potassium sulfite and ammonium sulfite. Hundreds of asthma sufferers have reported adverse reactions to the FDA after having eaten foods containing sulfites. Several deaths have been linked to acute reactions to sulfites, as reported by the FDA. In one such fatality, it was later found that samples taken from the meal included lettuce that contained 409 ppm of sulfite; the guacamole served at the same meal contained 272 ppm of sulfites.

SORBITAN MONOSTEARATE ∾ *See **Polyoxethylene (20) sorbitan monostearate.***

STEARIC ACID ∾ This substance is found naturally in some vegetable oils, cascarilla bark extract, and in tallow and other fats. Also known as *octadecenoic acid*, this compound is prepared synthetically through hydrogenation of cotton seed oil and other types of vegetable oils. It is the main ingredient in the making of soaps and lubricants. As a food additive, it is used in butter, vanilla flavorings, beverages, cakes, pies, and candy. The FDA sets limitation levels for this food additive at 4,000 ppm. One of the main concerns regarding this additive is its potential for raising the blood lipid levels of individuals who frequently consume foods containing this type of fat. About ten years ago, researchers from the University of Texas reported to the *New England Journal of Medicine* that stearic acid did not raise the blood cholesterol levels as much as some other saturated fats did.

SUCROSE POLYESTER ∞ *See* **Olestra®.**

SULFITE ∞ *See* **Sodium sulfite.**

THIAZOLE ∞ This chemical is incorporated in the synthesis of fungicides, dyes, and certain rubber products. It is sold under the names of *thiazole-4-YL, benzimidazole, thiabendazole,* and *arbotect.* A related fungicide, 2-(*thiazole-4YL*) *benzimidazole* (TBDZ) is used on animal feed, as an animal drug, and on citrus fruits and seed potatoes. The EPA's Genetic Toxicology Program has found that this chemical is poisonous if swallowed, and it has caused birth defects in laboratory studies.

THIODAN ∞ *See* **Endosulfan.**

TOXAPHENE ∞ This yellow, waxy, fragrant substance is employed as an insecticide in soybean oil. It is poisonous to humans if swallowed. The FDA tolerance levels for this chemical additive are 6 ppm in soybean oil. Lethal amounts of this chemical can penetrate the body through absorption by the skin, mouth, and lungs. In laboratory experiments with animals, toxaphene has been shown to produce birth defects.

VANICIDE ∞ *See* **Captan.**

XYLENOL (2-4 dimethyl phenol) ∞ This chemical is a coal tar derivative that is toxic when swallowed or absorbed via the skin. It is a crystalline solid that is used as a disinfectant and as a solvent in pharmaceuticals. In laboratory experiments, this chemical caused skin cancer in mice. Xylenol is also used as a saturating agent in the preparation of dyestuffs.

ZERANOL ∞ This chemical is a hormone that is employed for stimulating growth in cattle and sheep. However, the FDA limits residues for zeranol to zero in edible uncooked tissue of these meat sources. Therefore, if we can trust that a thorough job is being carried out in the final inspection of our meat products, we have no reason for concern.

NON-TOXIC ADDITIVES

AMBRETTE ∞ A natural flavoring agent derived from the seed of the hibiscus plant, ambrette is used in soft drinks, ice cream, candy, and baked goods.

AMMONIUM CASEINATE ∞ A protein naturally occurring in milk and cheese, ammonium caseinate is not required to be listed as an ingredient on the label of ingredients. It is used in baked goods.

AMYLASE (swine) ∞ An enzyme prepared from the pancreas of hogs, amylase is used in flour to break down starch particles into smaller sugar molecules in the preparation of dough. It improves the texture of baked goods and extends shelf life.

ANISE ❧ A nontoxic natural flavoring, anise is used in a variety of foods and beverages.

APRICOT KERNEL OIL ❧ This is a cold pressed, non-hydrogenated oil that can be used in salads.

ARACHIDONIC ACID ❧ Usually isolated from the liver of animals, arachidonic acid is an unsaturated fatty acid used for nutritional purposes and in skin creams and lotions for healing eczema.

ASCORBIC ACID (vitamin C) ❧ Because of its preservative and antioxidant qualities, ascorbic acid is added to frozen fruit, candy, artificially sweetened jellies, juices, and soft drinks. It is also used in the curing of beef or pork. It is an essential nutrient to human health. (*See* Chapter II.)

BAKING POWDER ❧ Certain brands of baking powder do not contain aluminum, which has been linked by some researchers to Alzheimer's disease. Aluminum-free baking powder is available at health food stores.

BAY LEAVES ❧ Extracts, oils, and oleoresins obtained from bay leaves are used to flavor soups, meats, baked goods fruit, liquor, desserts, and candy.

BETA-CAROTENE (provitamin A) ❧ Found in plant and animal tissue, this nutrient is the main source of natural color in butter, carrots, and egg yolks. It is used as a coloring agent in cosmetic products and in some foods. A vitamin A "precursor," it is stored in the body and transformed into vitamin A as the body requires it. (*See also* Chapter II.)

BIOFLAVONOIDS (vitamin P complex) ᕙ Citrus-flavored compounds derived from fruits and plants, bioflavonoids are used as a reducing agent (to remove oxygen). There are no regulations and no toxicity has been reported.

BREWER'S YEAST ᕙ First used in the brewing of beer, this microscopic plant contains many of the B complex vitamins, as well as protein. It is commonly used as a flavoring in vegetarian soups and as a nutrient. (*See* Chapter IV.)

BROMELAIN ᕙ Occurring naturally in pineapple, bromelain is an enzyme used in meats, meat-tenderizers, and beer. It is a strong protein-digesting agent.

CALCIUM SORBATE ᕙ A salt of sorbic acid, which is derived from mountain ash berries, calcium sorbate is a preservative and a fungus inhibitor used in baked goods, fruit products, desserts, beverages, and prepared salads. According to the Select Committee's report to the FDA, it is non-toxic, and there are no limitations on its use.

CALCIUM TARTRATE ᕙ Used as a preservative, it is derived from cream of tartar. It has no known toxicity.

CAPRYLIC ACID ᕙ Prepared from the oxidation of various essential and citrus oils, caprylic acid also occurs naturally as a fatty acid in the milk of cows and goats and in palm and coconut oil. It is used as a flavoring agent. It is often used in the manufacture of perfume. According to the FDA, this compound has no known toxicity; therefore, the FDA has set no limitations on it.

CASEIN ∽ This is the principal protein contained in cow's milk. It is used as a texturizer in ice cream, custard, and sherbets, and it provides the protein component in various shampoos and skin creams. It is non-toxic, and the FDA has set no limitations.

CHERVIL ∽ A natural flavoring extracted from an aromatic plant, it is used in beverages, bakery products, ice cream, candy, and condiments.

CHIVES ∽ A member of the onion family, chives are used to flavor many prepared foods and are believed to be non-toxic. The FDA has not yet made much information available regarding chives.

CHLOROPHYLL ∽ The substance that gives green color to plants, chlorophyll is essential to photosynthesis. It is used in oils, such as olive oil, and imparts a slight greenish tint. It is also used for its deodorizing abilities in such products as mouthwashes and deodorants. Excess amounts may cause photosensitivity in some individuals.

CHOLIC ACID ∽ This is a naturally occurring compound used as an emulsifying agent in dried egg whites and as a choleretic to regulate the secretion of bile. It is found in the bile of most vertebrates. No toxicity is associated with this additive, and the FDA sets no limitations on its use.

∽ ∽ ∽

CHOLINE ∽ Available alone or in combination with lecithin, choline is a part of the vitamin B complex and is essential to the metabolism of fat. It is widely used in poultry feed. The additive is usually taken from lecithin, which is isolated from egg yolks, soybeans, or corn.

CITRIC ACID ∽ Derived from citrus fruit through the process of fermentation, citric acid is a flavoring commonly used in beverages, ice cream, candy, and baked goods. It is also used as an additive to adjust acid-alkaline balance in many kinds of prepared foods. It is non-toxic and has no limitations, according to the FDA.

CRANBERRY JUICE CONCENTRATE ∽ This concentrate is derived from several plants belonging to the Vaccinium family. An extract of the red acidic berry can be used as a food coloring to replace synthetic red colorings that have been banned.

DULSE ∽ This reddish seaweed is added to foods in a powdered form as a flavoring. It has a slightly salty flavor, contains many essential trace minerals, and can be used in place of salt.

FENNEL SEED ∽ This herb is commonly used as a flavoring in many types of food products. It is used in the manufacture of prepared sausages because of its high anti-bacterial properties.

FOLIC ACID ∽ This nutrient occurs in nature and is an element of the vitamin B complex. It is found in liver, mushrooms, and dark green leafy foods. It is essential for the absorption of iron in the formation of healthy red blood cells. It is used commercially in the cosmetics industry.

GLUCOSE ∽ This compound is found naturally in corn sugars, grapes, and blood. It is an instant source of energy for plants, animals, and humans. Glucose syrup is added as a flavoring to other syrups, to sausage, and to other meat products. It is non-toxic, and can be used nutritionally.

GRAPESEED OIL ∽ This oil is used as an ingredient in fruit flavorings and has no known toxicity.

LECITHIN ∽ A natural emulsifier, emollient, and antioxidant derived from soybeans, corn, egg, and egg yolk, lecithin is used in breakfast cereals, baked goods, margarine, chocolate, desserts, and shortening. It has GRAS status, with no limitations other than to employ good manufacturing practices.

LEMON BALM ∽ Primarily an additive in creams, lotions, and perfumes, this herb can also be used as a food flavoring. It is non-toxic.

MALIC ACID ∽ This substance is found in many fruits, especially apples and cherries. It is used as a flavoring agent and aid for aging wines. It is also used in frozen dairy products, baked goods, fruit products, and other processed foods. Direct contact of malic acid on the skin can trigger an allergic reaction; however, as an additive this substance appears to be non-toxic, and the FDA has imposed no restrictions on its use.

NETTLES ❧ Derived from a bothersome weed equipped with protective stingers, nettles have a long history in folk medicine. A compound derived from nettles is currently used by farmers to prevent spoilage to tomatoes and as a growth stimulator for strawberries. It is considered to be non-toxic, but there is no information available from the FDA regarding toxicology or limitations.

PAPRIKA ❧ Commercial paprika is prepared by grinding the pods of dried, sweet peppers (*Capsicum annuum*). A good source of vitamin A, this reddish powder is used as a flavoring in many foods, including soups, meats, bakery products, and condiments. Paprika is also used as a food coloring. Limitations vary for this additive, depending upon its use. It has been used safely since the 1960s without incidence.

PECTIN ❧ This polysaccharide is obtained mainly from fruits, especially lemon and orange rind. It is used as a stabilizer, thickener, and a cementing agent in sherbet and ice cream, dressings, fruit preserves, and other food products. Some practitioners use it to lower blood lipid levels.

PHYTIC ACID ❧ This acid occurs naturally in cereal grains. Commercially, it is derived from corn. It is used today as a chelator (an agent used to remove heavy metals), as a treatment for hard water, and as an anti-rust compound. It is considered to be non-toxic.

PYRUVIC ACID ❧ This substance occurs naturally in coffee. It is also present in the body's muscles when sugar is being metabolized during exercise. A synthetic flavoring used in beverages, ice cream, and baked goods, pyruvic acid is extracted commercially from cane sugar. It is considered to be safe. The FDA will provide a full toxicology report of available information upon request.

RICE BRAN OIL ∞ Extracted from whole grain rice, this substance is used as a preservative and for coating candies. It is considered to be non-toxic.

SILICA ∞ A common element in nature, silica forms about one-eighth of the composition of all types of rock. It is used commercially in skin preparations for protecting the skin from harsh elements, and is also utilized as a coloring agent. The FDA considers this additive safe and has set no limitations for its use.

SORGHUM ∞ Although it is one of the most abundantly grown grains in the United States, less than 5% of the sorghum grown is used for human consumption. In developing regions, such as in some parts of Africa and Asia, however, it is primarily a food crop. Sorghum syrup is used as a sweetener and texturizer in food products. It is considered non-toxic.

TURMERIC ∞ The root of the East Indian herb *Curcuma longa*, turmeric is used as a flavoring or yellow food colorant in a variety of foods and beverages. Its extract and oleoresin are used in the manufacture of fruit, cheese, and spice flavorings. Turmeric has nutritive value and in the last several years was found by biochemists to contain medicinal and/or protective properties. Given the GRAS status by the FDA in the mid-1960s, turmeric is considered non-toxic.

VEGETABLE GUMS ∽ These gums are derived from quince seed, acacia, Irish moss, guar, potassium and ammonium alginates, and several other sources. Used in the cosmetics industry as thickeners, vegetable gums are considered by the FDA to be non-toxic, except for causing allergic symptoms in hypersensitive individuals.

WHEY ∽ This is the liquid that is left in milk after the extraction of fat and casein. Its most widely known use is in the making of cheeses. The FDA has given it the GRAS status and will supply up-to-date toxicology information upon request.

ZINGERONE ∽ This compound is available naturally in ginger. As a synthetic flavoring, it is added to such foods as sodas, candy, bakery products, and chewing gum.

Chapter II: Nutrients

While it is our Creator that many of us credit for the perfect piece of machinery that humans are, our biochemical individuality is a combination of heredity, environment, and the internal structure known as our "chemical makeup." We cannot control our inherited DNA patterns any more than we can select our parents, but we *can* look at our environment, nutritional choices, and life-style habits for a starting point we might be able to control. While together with the world's inhabitants we can effect positive changes in the quality of our air, water, and soil, we may have a more profound effect on "our own individual internal environment." Just thirty years ago, the medical establishment told us that the food we ate had nothing at all to do with our diagnosed diseases. Today, of course, we know that this just isn't so! We were not being deceived; the fact is, physicians simply did not know any better, and many know little more today regarding nutrition. Not much more than 3 hours (yes, you read me correctly!) is required in the study of nutrition for any medical degree.

What can we do to change this situation? Either we must demand that our health insurance carriers begin to pay for consultations with a qualified practitioner who is trained in the skills of nutrition, healthier lifestyles, and the herbs or we must accept that responsibility ourselves. This may be easier said than done for some, especially for those who are elderly or very ill already. Yet this type of service and care is essential to the health and future of all of us. Collectively, we can all make a huge difference in the way health maintenance and disease prevention is handled. Singularly, we can begin by taking back some of the responsibility and health care decisions that we so easily have given away. Also, we can begin to look to our environment for available and healthy nutritional choices. We can change many negative life-style habits shown to be associated with health problems, and, in that way, we may begin to exert some control over the state of our health, looking to tomorrow with confidence and with expectations for many positive health benefits.

We will also be better able to control eating disorders and overeating. Overeating is not always an emotional response. Often cravings are simply the body's signal to the brain for nourishment. When that cry is answered with foods that have little nutritional value, however, the feeling of being satisfied is temporary. This sets off a cycle of unrelenting, unanswered cravings for nourishment, which causes the individual to eat too much. A change in the diet will satisfy these cravings with less food, decreasing the problem of overeating.

This chapter covers vital nutrients that are of particular concern in maintaining a balanced diet and long-term health. (*See also* Chapter III, Dietary Sources, and Chapter IV, Supplements.)

∽ ∽ ∽

AMINO ACIDS

Amino acids are the chief components of proteins and are synthesized by living cells. Of the 22 amino acids that are used in protein synthesis, 8 must be obtained by human beings through diet. These are termed "essential amino acids." The body is comprised almost without exception of the 8 "L forms" of the amino acids, L-*tryptophan*, L-*isoleucine*, L-*lysine*, L-*threonine*, L-*leucine*, L-*methionine*, L-*phenylalanine* and L-*valine*. From these clearly described L forms, and some D-combination forms (not included herein), the body can manufacture raw elements into amino acids known as the "nonessential" amino acids. The nonessential amino acids are *alanine, arginine, asparagine, aspartic acid, carnitine, cysteine, cystine, glutamine, glutamic acid, glycine, histidine, hydroxyproline, proline, serine,* and *tyrosine.* When one considers using amino acids as a therapeutic supplement, in order to avoid over-dosing or triggering amino acid imbalances, one must ensure an adequate supply of all the essential amino acids. When taking amino acid supplements, it is strongly suggested that one be supervised by a health practitioner with substantial credentials.

Once again, I must remind readers that the information contained in this book is not meant to be a substitute for professional medical treatment when it appears to be necessary, nor is it meant that this information be seen as a path to "self-treatment!" What I offer to provide is a tool to become better educated regarding your body's chemistry and nutritional needs to help you make wiser choices regarding food selection.

Because of our unique chemistry and individual needs, specific requirements pertaining to the amount of these essential nutrients may vary. This can happen, in part, due to our excretion patterns. Two researchers who have led the way regarding the amino acids' role in nutrition are Roger Williams, and, more recently, Jeffrey Bland. Williams' book, *Biochemical Individuality*, offers detailed information on variations in nutritional needs. (Another reliable book that one might find in a local health food store is *Thorson's Guide to Amino Acids* by Leon Chaitow, a complete source of information on the essential amino acids.)

Protein-rich foods are the best sources of a rich supply of amino acids. The word "protein" is derived from a Greek root meaning "of first importance," and protein is the basic material of life. Protein constitutes three-fourths of our body tissue (excluding water composition). Muscles, organs, antibodies, some hormones, and all enzymes (the compounds that direct cellular chemical reactions) are largely composed of protein. Yet protein is not a single, simple substance, but a multitude of chemical combinations. The basic structure of protein is actually a chain of amino acids that can form many different configurations and can combine with other substances. The possible arrangements of the body's 22 amino acids are almost infinite, and thousands of different ones have been identified.

Proteins are continually being broken down in our bodies. Most of the amino acids are reused, but we must constantly replace some of those that are lost. This process is known as "protein turnover." Our need to keep this process going begins at conception and lasts throughout our lives.

Without sufficient dietary protein intake, growth and all bodily functions would not take place. Sources of the essential amino acids are milk, eggs, cheese, all meats, seafood, soybeans, tofu, and many of the nuts and seeds. While fruits, vegetables, and grains may be rich in protein, they rarely contain all 8 essential amino acids in one source. Many vegetarians who are not being provided the full amount of essential amino acids regularly can often be putting their health at serious risk. Since protein is the main source of the amino acids, vegetarians need to understand how to maintain an adequate supply. Food combining (rice and corn, rice and beans, protein drinks, and soy products) are ways to ensure an adequate supply of essential amino acids.

Protein rich food should comprise at least 40% of each meal. If supplementation is desired, it is recommended that one take a supplement that contains a good source of all of the essential amino acids, lest one be susceptible to creating an imbalance. Individuals with suspected genetic deficiencies should visit a professional who specializes in such disorders.

The following entries provide brief summaries of essential information on each amino acid and the role it plays with regard to our health.

ARGININE ∽ An essential amino acid for children but not for adults, arginine is secreted by the anterior pituitary gland, which in turn stimulates *human growth hormone* (HGH). HGH helps the wound-healing process. Arginine is absolutely necessary for normal sperm count, stimulates fat metabolism, and is involved in glucose metabolizing and insulin production. Deficiencies are linked to infertility in males, premature aging, obesity, and increased free radical activity. Overabundance of this amino acid has been associated with schizophrenia and abundant virus replication of herpes simplex, unless sufficient lysine is available. Arginine supplementation is used therapeutically for arthritis, tumor inhibition, and stimulation of T-cells, which are involved in the body's immune function.

ASPARTIC ACID ∽ This nonessential amino acid is responsible in part for the transportation of magnesium and potassium to cells. Aspartic acid also is an element of the sweetener aspartame (along with *phenylalanine*). It is beneficial in the energy cycle and is useful in promoting the metabolism of trace minerals. It helps the liver detoxify ammonia. It has been therapeutically useful in relieving patients supplemented with potassium and magnesium aspartate.

CARNITINE ∽ Considered a nonessential amino acid, carnitine is needed in larger amounts by men than by women, and it is notable for its effects on sperm motility. Its conversion in the liver through the presence of other available amino acids is dependent upon vitamin C. Positive effects include mobilizing fatty deposits in the obese and removing lipids from the bloodstream. It is of therapeutic benefit for some forms of infertility, circulatory problems, relieving fatty deposits in the liver, and in a few other disorders. **Cautions:** Individuals with kidney disorders should avoid supplementing their diet with this amino acid. A majority of patients who have used carnitine have complained of gastrointestinal side effects and/or an increased body odor, which eventually disappeared when the initial dosage was decreased.

CYSTEINE ∽ A nonessential amino acid, cysteine is synthesized in the liver from *methionine*. It should not be confused with *cystine*, which does not provide the antioxidant benefits that cysteine does. In the absence of adequate vitamin C, cysteine converts itself into cystine. It is beneficial for the strengthening of the hair, and is also involved in the process of enzyme and insulin conversion. Because of its free radical combating powers, cysteine also improves skin texture and flexibility. Cysteine can be useful as a supplement in nearly all cases of chronic disease. As a chelator, cysteine binds with heavy metals and helps in their elimination from the body. It is useful in protecting the body from toxins from pollution, alcohol, and other sources. **Caution:** This amino acid should be used by diabetics only with extreme caution.

CYSTINE ∽ A nonessential, sulfur-rich amino acid, cystine also behaves in the body as a heavy metal chelator. Due to its rich sulfur content, it has been utilized in the treatment of skin problems. **Caution:** Individuals susceptible to kidney or liver stones may be wise to avoid supplementation of this amino acid.

∽ ∽ ∽

GAMMA-AMINOBUTYRIC ACID (GABA) ∞ A nonessential amino acid synthesized from *glutamic acid*, GABA assists in regulating nerve function and stimulates the performance of *vitamin* B₃ (*niacinamide*). When used therapeutically it has been shown to produce calmness and tranquility in patients diagnosed with manic disorder, schizophrenia, and high blood pressure. It has also been utilized for patients with enlarged prostate glands, accomplishing reduction of the enlarged prostate gland, it seems, through its stimulatory effect and release of the hormone *prolactin*.

GLUTAMIC ACID AND GLUTAMINE ∞ These two chemicals vary only slightly in their structures, and both are considered to be nonessential amino acids. Under certain circumstances, however, they can be considered essential amino acids. These two amino acids are brain fuels, components in the glucose tolerance factor, and required for the production of other nonessential amino acids. These are the most abundant amino acids in the cerebral spinal fluid. In the absence of these amino acids, *folic acid*, an essential B vitamin, cannot be manufactured. These aminos can easily convert themselves from one to the other, as needed by the body. For instance, glutamic acid converts itself into glutamine for the flushing of ammonia from the brain. Some laboratory experiments point to the fact that glutamic acid may be able to dissolve stone formations from the kidneys. Glutamine has been shown to be useful in some types of depression, senility, and schizophrenia. Furthermore, it is said to lessen one's desire for carbohydrates and refined white sugars, perhaps being beneficial for those interested in managing their weight. A deficiency of this particular amino acid is said to promote an irritable or rebellious attitude. **Caution:** Toxicity can result from doses over 2 grams daily and may cause manic behavior.

GLUTATHIONE ∽ This nonessential amino acid comprises *cysteine, glutamic acid,* and *glycine.* It is beneficial in slowing the aging process, neutralizing toxic atmospheric chemicals, and sparing damage to fat cells caused by free radicals. Used therapeutically, it protects the liver from damage caused by alcohol, works as a heavy metal chelator, and has shown an ability to retard tumor growth in animals.

GLYCINE ∽ A nonessential amino acid known as part of the tripeptide glutathione, glycine is required for the synthesis of bile acids and is used as a sweetener. This amino acid is probably best known for its detoxifying effect on the liver. It is not used alone therapeutically .

HISTIDINE ∽ This amino acid is required for sexual arousal. It is metabolized into the neurotransmitter *histamine*. Histamine is related to smooth muscle performance, along with contraction and dilation of blood vessels. It helps maintain the healthy condition of the myelin sheaths (coverings which help protect the nerves) and has also proven therapeutic for some types of schizophrenia—those which are characterized by their imbalances of histamine levels. Deficiencies of this amino acid can lead to impaired hearing. **Caution:** women with premenstrual depression or patients diagnosed as manic-depressive with higher than normal levels of this amino acid should avoid supplementation.

ISOLEUCINE ∾ *See Leucine.*

LEUCINE AND ISOLEUCINE ∾ Combined with the amino acid *valine*, these two essential amino acids form a family known as *branched chain amino acids*. These two amino acids are often most deficient in chronically ill patients, according to documented profiles. Because they work best as a pair, they should ordinarily be supplemented together, unless an experienced professional is implementing a customized therapeutic protocol with a targeted result. The amino acid leucine has lent itself favorably to the treatment of Parkinson's disease, and D-leucine may be useful for eliminating or controlling pain. **Caution:** an overabundance of these amino acids may predispose individuals to pellagra (a skin condition caused by a nutritional deficiency).

LYSINE ∾ An essential amino acid, lysine is required for the metabolism of fats. It is often deficient or low in vegetarians. Lysine assists the body in the formation of antibodies needed to fight disease and is required by children for their proper growth and development. Deficiencies of this amino acid result in dizziness, fatigue, anemia, nausea, and eyesight problems. A deficiency of long standing could make a person vulnerable to an increased excretion of calcium and susceptible to kidney stones. Used therapeutically, it has been shown to reduce the risk of atherosclerotic changes (hardening of the arteries) and possibly enhance one's concentration abilities.

METHIONINE ∾ This essential amino acid is assisted by vitamin B$_6$. Methionine contains sulfur and is a strong antioxidant that helps in the prevention of free radicals.

A precursor of a few of the other amino acids, methionine assists in the production of choline, vitamin B_{12}, and lecithin. When used therapeutically, it is recognized for its detoxification abilities. It assists the body's attempts to chelate toxic metals from the body and remove any overabundance of histamine. It is essential for the availability of selenium in the body and also assists in the detoxification of the liver. A deficiency of this amino can result in poor skin tone, hair loss, anemia, atherosclerosis, toxicity, and fatty liver deposits. **Caution:** There are disorders that rule out the supplementation of this particular amino acid. Postmenopausal women have been shown to be more inefficient in metabolizing methionine, and some showed high levels of homocystein afterward in some clinical findings. Pregnant women and those anticipating pregnancy might not be good candidates for methionine supplementation.

ORNITHINE ∽ Working together with *arginine*, ornithine is beneficial in the removal of ammonia during the urea cycle (the main breakdown product of protein metabolism). It is a strong stimulator of growth hormone and increases body metabolism. It may benefit therapeutically a weakened immune system and those suffering from autoimmune disorders such as rheumatoid arthritis.

PHENYLALANINE ∽ This essential amino acid, with vitamins B_6 and C, helps to stimulate the manufacture of *tyrosine* and *norepinephrine*, compounds that are necessary for the management of the heart rate, oxygen consumption, blood sugar levels, blood pressure, and metabolism of fat. Phenylalanine cannot be metabolized in the absence of sufficient amounts of vitamin C. The thyroid requires this amino acid for normal functioning. The L-form stimulates production of *cholecystokinin*, inducing a feeling of having eaten enough, and can therefore be useful in weight management. In its D-form, it reduces symptoms of multiple sclerosis and Parkinson's disease, and may help to alleviate depression and improve memory. In conjunction with ultraviolet light, it has been utilized to treat the skin disease known as *vitiligo*. Deficiency symptoms can cause a tyrosine imbalance, which in turn can lead to mental retardation or eczema in children, as well as weight gain, circulatory problems, and emotional disorders in adults. **Caution**: Individuals who are taking any of the *monoamine oxidase inhibitor* drugs (MAOs), or who have any type of high blood pressure problems, should not consider supplementation with this amino acid. Also, it is best avoided by pregnant or lactating women or those with genetic protein metabolism problems.

PROLINE ∽ Necessary for optimum skin health, proline, a nonessential amino acid, is also a necessary element in the forming of muscle and connective tissue, along with vitamin C and several essential minerals.

TAURINE ∽ Taurine is manufactured in the body from another nonessential amino acid, *cysteine*, the major supply of which should come from the diet and is found only in foods of animal origin. Amounts of taurine are regulated by the availability of zinc and can, therefore, fluctuate. Taurine has been found to be inhibited in the liver by the presence of the female hormone *estradiol*; therefore, it is understandably required more by women than by men. While research is ongoing regarding this amino acid's connection to heart function, it is regarded as extremely important. More prevalent in the heart than any other of the amino acids, it is potassium and calcium sparing. Taurine, only found in the L-form, is also involved in the regulation of insulin and blood sugar levels. An individual's need increases when under stress. It is the second most-prevalent amino acid in human milk; little is supplied in cow's milk. It has been employed in the treatment of congestive heart failure, immune functioning, and atherosclerosis. Deficiencies in zinc and this amino acid may lead to eye problems in adults and epilepsy in children.

THREONINE ∞ This amino acid is necessary for mental clarity and proper digestion. Because it is plentiful in pulses (peas, seeds, lentils and beans), it is a major food selection for vegetarians in order to fulfill the complete amino acids requirements.

TRYPTOPHAN ∞ An essential amino acid, tryptophan has many uses within the body. It is a precursor of the neurotransmitter serotonin, one of nature's balancers, which is a calming substance affecting mood and sleep patterns. It can also affect particular food choices and can therefore become an aid in the management of weight gain. The uptake of tryptophan by the brain is enhanced by vitamin B_6 and vitamin C. Other therapeutic values range from alleviating the onset of migraine headaches to treating restless leg syndrome. Also, it has been recommended to patients with tardive dyskinesia (a central nervous system disorder characterized by involuntary motor movements) and arthritis.

Caution: Tryptophan is not to be used by pregnant women, individuals diagnosed with lupus, or anyone with asthma.

TYROSINE ∞ This nonessential amino acid can be manufactured by the body from phenylalanine. It is useful for stimulating brain neurotransmitters, as long as it is taken in small doses. Grass allergies and hay fever have responded positively to tyrosine.

VALINE ∞ An essential amino acid and part of a combination referred to as *branched chain amino acids* valine is necessary for neural, mental, and muscle functioning. It has also been said to be beneficial in reducing inflammation. Hallucinations and a sensation of "crawling skin" have been associated with an excess of this amino acid, while deficiencies have been linked to toxicity, nervousness, and interrupted sleeping patterns. In addition, certain genetic disorders involve imbalances of this essential amino acid.

FATTY ACIDS

There is much confusion regarding the subject of fats. Many assume that all fat is bad and must be eliminated from the diet, although in reality some fat is absolutely essential. What is important is the type of fat that is consumed. The essential fatty acids (EFAs), also known as *linolenic* and *linoleic* acids, or the *unsaturated* fatty acids, are the "good" fatty acids, sometimes referred to as vitamin F. Although these essential fatty acids are involved in many vital functions, the body is not able to manufacture them; therefore, they must be supplemented. Following is a summary description of the various types of fats:

HYDROGENATED FATS ⁓ When vegetable oils or shortenings are heated at very high temperatures, their unsaturated fats are transformed into a more solid or saturated form of fat, in a process known as "hydrogenation." Unsaturated fats (or the "good fats") are destroyed when huge amounts of hydrogenated fats are consumed. Hydrogenated fats may inhibit the body's ability to fight free-radicals, thus increasing the risk for cancer. Hydrogenated oils (those which are mostly solid at room temperatures, such as margarines) can contribute to high cholesterol, and coronary heart disease.

MONOUNSATURATED FATS ⁓ These are fats that are derived only from vegetable sources. They include avocados, canola oil, cashews, almonds, hazelnuts, pecans, and olive oil in soft or solid form. They are often the types of oil most often recommended in the diet of individuals who suffer from hyperlipidemia (high cholesterol), as these oils do not effect the blood lipid levels.

POLYUNSATURATED FATS ⁓ Polyunsaturated fats is simply another name for the essential fatty acids—*linoleic* and *linolenic* acids, or vitamin F. (These essential fatty acids are covered in more depth after this introductory discussion.) These fatty acids are the "good fats" that can actually lower the "bad" cholesterol, while raising the "good" cholesterol, or HDL. These include cod liver oil, corn oil, flax oil (which is the highest, purest form of the unsaturated fatty acids), primrose oil, safflower oil, soy oil, sunflower seeds, sesame oil, and wheat germ oil.

SATURATED FATS ∽ These are the culprits which lead to clogged arteries, and, therefore, should be avoided or significantly limited in the diet. In fact, the American Heart Association suggests that the process of fat collection within the arteries begins very early on in life, and that children as young as two years of age should have their fat intake limited to prevent this. Saturated fats are found in all animal products, as well as in many vegetable oils. Common sources of saturated fats include bacon, butter, beef, processed cheeses, chocolate, coconut and coconut oils, milk and cream, palm kernel oils, pork, and poultry (especially the skin). These electron-deficient fats inhibit the electron exchange necessary to living tissue, because they, like tar, behave as insulators against electrical conductivity, and thus they weaken the life functions in organs and various growth centers throughout the entire body. Also, like tar, they are carcinogens.

UNSATURATED FATS ∽ This is another term for the "good fats," including the essential fatty acids, linolenic and linoleic acids, vitamin F, polyunsaturated fats, or unsaturated fats.

ESSENTIAL FATTY ACIDS

The unsaturated, electron-rich essential fatty acids (EFAs) include linoleic and linolenic acids, unsaturated oils, natural aromatics from herbs, spices, and fruits (which are abundant in aromatics and natural color components corresponding to the photons of sunlight). The EFAs increase the absorption, storage, and utilization of the sun's energy in a process which is very similar (within the human body) to the type of photosynthesis most of us are familiar with from studying plants. All of this information was made known, utilized therapeutically, and passed down to us from Dr. Johanna Budwig, as far back as the mid-1950s. A world renown biochemist in Europe, Dr. Budwig was a seven-time Nobel Prize nominee and a foremost authority on fats and healing.

Some of the benefits derived from including a daily supplement of the essential fatty acids include:

(1) A softening of tumors. As an experiment, place a hard piece of suet into a container and cover it with an essential fatty acid (organic flax oil is the highest, purest form, according to Dr. Budwig). As the days begin to unfold, you will be able to witness the disintegration of the suet into smaller, softer molecules, which the body can then easily excrete.

(2) Nutrient utilization. EFAs enable the body to better extract and utilize the nutrients (or energy) contained in the fresh fruits and vegetables we eat.

(3) Circulation system benefits. EFAs provide a suppleness and elasticity to the arteries, and, thus, aid in the prevention of arteriosclerosis (hardening of the arteries).

(4) Heart disease prevention. EFAs raise the good (HDL) cholesterol and lower the bad (LDL) cholesterol, thereby helping to prevent heart disease.

(5) Beauty aids. The EFAs are an essential nutrient, required for optimum health of the skin, hair, and nails.

(6) Brain communication. The EFAs stimulate the production of neurotransmitters, which are necessary for carrying messages from the brain to various parts of the body.

(7) Allergy aid. The essential fatty acids help to reduce the amount of histamine that is released by the body in response to various allergens.

(8) Mucus production. The relationship between unsaturated fats and proteins is extremely important in glandular functioning, because glands cannot produce mucus if the unsaturated fats are missing!

(9) Other benefits. Further study may reveal several other positive reasons for making the unsaturated fats an important part of one's daily fat intake; however, what has been included here should provide sufficient motivation to include the EFAs into your daily caloric intake.

The purest form of the essential fatty acids is flax oil. It is a cold-pressed oil that can oxidate if heated and should be kept refrigerated. It is not meant to be a cooking oil. Dr. Johanna Budwig claims to have realized extremely positive results with flax oil in the treatment of certain cancers. Flax oil also increases circulation, thereby enhancing oxygen utilization, improving energy, enhancing metabolism, nourishing skin, reducing allergic responses, stimulating and strengthening the heart and lymph systems. The list of the benefits of EFAs goes on.

Recommendations: The essential fatty acids (or vitamin F, linoleic and linolenic acids) are totally non-toxic, and suggested recommendations are 3% of total calories for infants and 2% for adults, although the National Research Council has not yet set dietary allowances for these fats. The essential fatty acid needs are met when the diet includes calories produced by linoleic acid, found in foods such as vegetable oils, including soy, flax, corn, and sunflower, as well as wheat germ. **Caution:** there are no known toxic effects; however, excessive amounts might contribute to a possible metabolic disturbance and/or weight gain. One tablespoonful of organic flax oil combined with 1/4 to 1/2 cup of unprocessed cottage cheese or yogurt (a complete protein form) daily will meet the needs of most individuals.

MINERALS

The minerals are most associated with the "total substance of man." In The New World College edition of *Webster's Dictionary*, substance is defined as "the real or essential part or element of anything; essence, reality, or basic matter, particular chemical composition, the real content, resources, or wealth." Without adequate amounts of each of these essential minerals, the skeletal frame that we depend upon for the simplest task of walking would fail us. With insufficient calcium or copper, our hearts would become weakened. Also, without minute amounts of essential trace minerals, certain specific hormone reactions could fail to occur. With an excessive accumulation of toxic metals in our organs, however, we could literally "lose our minds," or succumb to death itself!

There are 18 essential minerals and approximately 15 trace minerals, which constitute the body's total mineral content. Although some minerals may be required only in minute amounts, this fact does not lessen their importance to the body's proper performance. The minerals that we seem to have no apparent need for may be present in minute quantities to help establish another mineral's ability to remain stable, or available, and to be better utilized by the body. Future studies may discover more reasons and uses for these minerals in small quantities in humans. The minerals are listed alphabetically.

ANTIMONY ∽ A metalloid element, antimony is a brittle, crystalline substance used in combination with other alloys in metals to harden them and increase their resistance to chemical action. Compounds of antimony have been utilized in medicines, pigments, and in matches for the purpose of "fire-proofing." While various levels of this element may be present in certain individuals, its need in the human body has not been determined, and high levels could pose a potential problem.

BARIUM ∽ This silvery-white, metallic element is somewhat pliable and can be found as a sulfate or carbonate. Barium has no known benefit to the human body; however, excessive levels have been linked to certain specific cardiovascular diseases.

BISMUTH ∽ This solid, brittle metallic element is grayish-white but often displays a faint red hue. While it has been employed in the manufacture of medicines, it is not determined to be of any benefit to the human body.

BORON ∽ Isolated in 1808 by chemist Sir Humphrey Davy (who gave this mineral its name), this metalloid element is found in combination with sodium and oxygen. Its chemical consistency is that of one or two distinct compounds, a brownish powder or a solid brilliant crystalline substance. Widely used today in the manufacture of pottery, glass, enamel, soaps, and boric acid, boron is considered an essential mineral. Among other recently revealed discoveries, boron works in conjunction with other vitamins and minerals (vitamin C, vitamin D, and calcium) to build and maintain healthy bones. Deficiencies of boron can lead to incomplete absorption of calcium and possible bone disease, including arthritis. Boron is also said to increase muscle and brain function, but can become toxic in high doses; in fact, ingestion of more than 3 milligrams daily is not recommended. Toxic symptoms include skin disorders, bowel disturbances, loss of energy, and diarrhea. Natural sources of boron include fruits, vegetables, and nuts.

CALCIUM ∞ One of the most familiar minerals known, calcium was also discovered in 1808 by Sir Humphrey Davy. A soft-white metalloid element, calcium is always found in combination in compounds such as marble, chalk, and limestone. Although used commercially in fertilizers, calcium is absolutely essential for human life on many levels. It works in combination with vitamin K (the clotting factor in the blood) to prevent hemorrhaging, and it is also vital for the health and performance of the heart, nervous systems, and muscles. Calcium has been used therapeutically to lower blood pressure, to assist the heart in maintaining a steady heartbeat, and to excrete excess sodium. Insufficient calcium may be associated with irregular heartbeat, depression, and insomnia. Without calcium, our teeth would become soft and porous.

Chronic deficiencies could precipitate symptoms associated with nerve and/or muscle (including the heart) malfunctions. Osteoporosis is characterized by an inadequate supply of calcium, phosphorus, and vitamin D. A rare condition known as *hypocalcemia* is possible when the blood calcium level drops below 8 milligrams. Chronic fatigue, renal failure, and nervous system dysfunction may be caused by extremely low levels of calcium in the body tissues. Excessive calcium, especially inorganic forms, might not be properly utilized by the body. It may travel to isolated places in the body, where it could begin to calcify, or become hardened and useless. This results in a benign tumor, which can, however, transform at some point (depending upon the general health of an individual) into a cancerous one. Excess calcium can also inhibit the performance of both the thyroid and the adrenal glands, causing a lack of energy. An extremely high intake of calcium along with vitamin D is linked to a disorder known as *hypercalcemia*, which denotes calcification of the bones, some of the tissues, or the kidneys. Whenever excess amounts of calcium are added to the blood plasma, coagulation will not occur and can, therefore, lead to excessive bleeding.

As is the case with most of the other essential minerals within the human body, a constant state of chemical *balance* is necessary in order to maintain proper functioning. Calcium works together with magnesium, phosphorus, sodium, potassium, and lead. The proper ratio among these minerals will have a great influence upon how the internal system performs. For example, excess calcium consumption can contribute to a deficiency in magnesium, or phosphorus.

Also, excess levels of phosphorus, or magnesium can lead to a calcium deficiency. Calcium absorption is related to vitamin D (calcitonin) availability. When this internal chemistry is upset, or becomes unbalanced, an excessive buildup of one particular mineral will either stimulate excessive excretion or inhibit absorption of another essential mineral, which in turn can impair the optimum performance of one or more of the body's internal organs dependent upon that particular mineral. In the past, many physicians have recommended calcium supplementation in the treatment of certain bone disorders. We have since learned that the form of calcium recommended (preferably citrate, which is the most natural and compatible type accepted and best utilized by the human body) as well as the minerals taken with it may be more important than the actual amount of calcium ingested. For proper assimilation, calcium requires vitamin D, phosphorus, and magnesium. The National Research Council recommends 800 milligrams daily of calcium intake, a high amount because the body can only fully absorb about 20 to 30% of that amount (240 milligrams). During pregnancy or for lactating women, that recommended dosage increases to 1,200 milligrams. Magnesium levels are available in predetermined proportions to calcium; magnesium's intake must be increased accordingly. In other words, 600 milligrams of calcium is proportionate to 1,000 milligrams of magnesium.

Natural sources of calcium include millet flour, organic almonds, bananas, yogurt, soy, oats, dark green leafy vegetables, and brewer's yeast.

∽ ∽ ∽

CHROMIUM ∞ The metallic element chromium has been gaining much popularity and recognition among naturopaths and the more health-conscious minded individuals. Chromium was discovered in 1797 by the French chemist Vauquelin. So named because of its brilliance, this grayish-white colored element is a hard crystalline compound. It is found in chemical substances containing nickel, copper, manganese, and several other metals. Considered to be an essential mineral, chromium is involved in assisting the body to activate certain needed enzymes and to help transform glucose into available energy. In the blood chromium is found in concentrations of 20 parts chromium to 1 billion parts blood. With certain amino acids and niacin, chromium is an integral part of an active substance known as Glucose Tolerance Factor, or GTF. It is also involved in the process of transforming fatty acids and cholesterol into a form which makes it easier for the body to excrete, and is most beneficial to the diabetic. Many use chromium supplements for such recognized benefits as memory enhancement, weight loss, combatting fatigue, and for its known glucose sparing effect. Chromium has been used therapeutically to control unstable sugar levels in the blood and for newborns diagnosed with "kwashiorkor." The main route of excretion of chromium is in urine; therefore, daily requirements are ascertained according to the amount of urine excreted. There are often inconsistencies in the proper recommended intake, probably due to factors such as one's geographic location and that location's soil content, one's genetic makeup, or one's medical diagnosis. The United States seems to be the country with the largest depletion of chromium in the soil.

When even a slight deficiency exists, the body will begin to show symptoms. A daily intake could be anywhere from 50 to 200 micrograms. Also, the amount of chromium stored in the body appears to fall as we begin to age. Chromium is found most abundantly in whole grain cereals and breads, dried beans, corn oil, potatoes, brown rice, cheese, and brewer's yeast.

COBALT ∾ This hard, shiny, steel gray metallic element, once thought to be of no value by miners, was renamed in 1730 by Swedish chemist George Brandt. It has been utilized throughout the centuries in everything from inks and paints to radioactive isotopes to the treatment of cancer. An essential mineral and an integral part of vitamin B_{12}, or *cobalamin*, cobalt is able to activate the production of certain enzymes within the body, especially those necessary for the health and maintenance of blood cells. A deficiency of cobalt can lead to some types of anemia or slowed growth rate. Excess cobalt is stored in the liver, kidneys, pancreas, and spleen. Toxic effects attributed to high doses include an enlarged thyroid gland, heart palpitations, diarrhea, and numbness in the fingers or toes. There is no recommended daily allowance (RDA) provided for cobalt, but because the dietary need is low, it is believed that 6 to 7 micrograms is sufficient for a daily intake. The best sources of cobalt are liver, kidney, oysters, clams, meats, and milk; thus vegetarians could be susceptible to a cobalt deficiency, which could, if undetected, contribute to an irreversible nervous disorder.

∾ ∾ ∾

COPPER ∞ Found in minute amounts throughout most body tissues, copper is considered to be an essential trace mineral. It is found in temperate regions. This somewhat pliable, reddish-brown element is most noted for being a corrosion inhibitor and an excellent conductor of electricity. Copper is found in many enzymes, and it assists in the vital formation of hemoglobin and red blood cells by facilitating iron absorption. Additionally, copper is necessary for the strength of *phospholipids*, those chemical compounds vital to the formation of the myelin sheaths, the thin tissue coverings that surround and protect many of our internal nerve fibers and endings. Without sufficient copper in the body, we might be more susceptible to coronary heart disease, and iron might be incorporated less into the hemoglobin. This is true because copper and iron are considered to be compatibles. (Compatibles are two are more of any substances that compliment the others. If one is found separated from its compatible or surrounded by an "incompatible" or opposing substance, ideal performances or optimum results are negated.) Some minerals in excess are associated with specific metabolic types. Phosphorus, sodium, iron, selenium, and manganese are more accurately associated with the "fast metabolizer," while the elements of calcium, copper, chromium, magnesium, and zinc best describe the "slow metabolizer." Excessive amounts of copper in our body's tissues make us more vulnerable to constipation, breathing difficulties, anemia, aneurysms, arthritis, heart disease, and premenstrual syndrome (PMS). The National Research Council recommends a daily intake of 2 milligrams for adults, but other experts state that between 2 and 3 milligrams is ideal. Copper intake over 40 milligrams can have toxic effects. Natural sources for copper include seafood, egg yolks, legumes, whole grains, soybeans, raisins, oats, nuts, almonds, and pecans.

GERMANIUM ∾ Germanium is a grayish-white metalloid element belonging to the carbon family and used in transistors, infrared equipment, and other similar products. It is often found in variable amounts in the human body, but its essential requirement has not yet been determined; therefore, no recommendation for germanium is offered. Although its presence and/or absence in the human body's delicate chemistry is presently difficult to evaluate, some experts state that high intake and absorption of this element has been linked to certain kidney disorders.

GOLD ∾ This is a pliable, ductile metallic element. Gold is considered to be one of the precious metals, often used in jewelry and coins and associated with money, brilliance, wealth, and status. Gold has no known essential value to the human body. Once utilized therapeutically for the treatment of skin and kidney disorders and arthritis, this practice came to an end when patients who were being treated with injections of gold began to experience toxic effects.

IODINE ∾ Belonging to the halogen family, this nonmetallic element comprises gray to black crystals. It has been used as an antiseptic and in the manufacture of dyes used in photography. A radioactive isotope of iodine is used today for the treatment of thyroid disorders. Deficiencies of iodine are not usual, and this is attributed to the fact that the body's essential requirement for iodine is very low. Most deficiencies are due to a specific problem often attributed to an individual's inability to assimilate or store this trace mineral. Symptoms often associated with an iodine deficiency include a noticeable decrease in energy, hypothyroidism, nervousness, excess weight gain, irritability, or skin and hair problems. Natural sources of iodine include table salt (sodium chloride), seafood, spinach, onions, garlic, carrots, tomatoes, pineapple, and kelp. Because sodium chloride (table salt) has been linked to hypertension and possibly to stomach disorders, kelp (or sea salt) is the preferred source of most health conscious individuals and practitioners. An added benefit is that it also include many of the essential trace minerals required for optimum functioning, although excessive doses can trigger headaches. In such cases, simply lower consumption and continue use.

IRON ∞ In earlier times, iron was simultaneously associated with such terms as holy, strong and/or vigorous movement. It is a malleable, white metallic element that is easily magnetized and quickly rusts in moist or salty air. Iron is truly one of the strongest metals ever to be discovered and has had a significant impact on the lives of humans. Iron is also essential in the body, where it works in several ways. It is responsible, in part, for removing energy from the "citric acid cycle" and in the manufacture of blood cells. Deficiencies of iron can lead to fatigue and anemia. In some instances, imbalances of iron have been found to be associated with eating disorders, including bizarre behavior.

Among these disorders are *pica*, and *amylophagia*. Other symptoms of a deficiency of iron include lack of energy, headaches, heart palpitations during exertion, a pronounced vulnerability to infections, dizziness, reduced immune response, and reduced white cell count. Avoid iron supplementation, unless recommended by one's practitioner. Too much iron (100 milligrams) can produce toxic effects. Iron overload often effects men more than women. A serious iron accumulation causes damage to the liver, and iron deposits have been found in the heart, joints, lymph nodes, skin, and even in the brain. Alcohol is known to increase iron absorption. Some researchers have found that an imbalance of iron to copper may play a role in Parkinson's disease, as well as in other neurological disorders. In fact, excess accumulations of specific minerals may one day allow us to predict such personality traits as irritability, impatience, and aggression.

There are plenty of sources of iron in today's menus, and the problem with most deficiencies can often be traced to genetic disorders and/or a deficiency in other nutrients necessary for the absorption of iron, such as vitamin B_{12}, vitamin C, folic acid, and hydrochloric acid. The National Research Council recommends a daily iron intake of 18 milligrams for women and 10 milligrams for men. Natural sources of iron include whole grains enriched with iron, meat, fish, poultry, eggs, leafy green vegetables, potatoes, fruit, figs, raisins, blackstrap molasses, beets, asparagus, and soybeans.

∾ ∾ ∾

LITHIUM ∾ Discovered in 1818 by A. Arfwedson, a student of Berzelius, this soft, silvery white metallic element, the lightest known metal, is now employed in the manufacture of thermonuclear explosives. Little is known regarding lithium at this time, including exactly how it works in the human body, but most of us store and utilize small amounts. It has been used with some degree of success for the treatment of certain mental/emotional disorders, such as bipolar disorder (manic-depression). Excess lithium may interfere with the uptake and/or balance of certain other essential minerals. One example of excess lithium is a condition known as *hypothyroidism*, which results from lithium blocking the thyroid's ability to balance the *parathyroid's* activity.

MAGNESIUM ∾ A silver to white metallic element, magnesium is utilized in alloys. Magnesium is considered an essential mineral; it makes up about 0.05% of the body's weight. Almost three-quarters of the body's magnesium content can be found in the bones, along with calcium and phosphorus. The remaining 25 to 30% is taken up in the soft tissues and absorbed into the body fluids. Magnesium helps the body absorb nutrients, and it assists enzyme function and metabolism (along with other minerals, vitamins, and nutrients). Required for the health and function of nerves and muscles, including the heart, magnesium is also known as a compatible element, often referred to as a synergist. Deficiencies of magnesium may initiate such symptoms as tremors, convulsions, insomnia, heart and behavioral disturbances, bruxism (teeth grinding), memory

loss, rapid pulse, anxiety, hyperactivity, fatigue, and confusion. A deficiency of magnesium can easily occur, because this mineral is refined out of many foods during
processing and can be destroyed in cooking. Magnesium
deficiencies can occur in individuals with pancreatitis, diabetes, chronic alcoholism, including cirrhosis of the liver,
and chronic diarrhea. When hormones are employed as
drugs, they too can cause a metabolic deficiency of magnesium. Toxicity caused by excessive magnesium intake is
rare; however, "hyper magnesia" can occur, often a result of
a decrease of urinary output. The presence of certain bone
tumors can also increase the absorption of magnesium.
Toxic symptoms can lead to depression of the central nervous system and, in extreme cases, death. Magnesium
must always be taken in combination with calcium (unless
specifically recommended otherwise by a skilled practitioner), and the National Research Council recommends a
daily minimum of 350 milligrams for the adult male and
300 milligrams for the adult female, 450 milligrams during
pregnancy and in lactation. Natural sources of magnesium include whole grains and vegetables, apples, soybeans, and brewer's yeast.

MANGANESE ∾ Manganese is a normally hard, brittle, grayish-white metallic element, similar to iron in that it rusts
easily, but unlike iron it is not magnetic. It is an essential
trace mineral, vital to normal bone development, and it may
also play an important role in the formation of blood. It is
also required for the production of *urea*, an important factor
of the urine, and it assists in sustaining adequate levels of

specific sex hormones. Due to its ability to metabolize fatty acids and cholesterol, manganese has proven to be of benefit to diabetics. Manganese (in combination with B vitamins) has helped individuals suffering from devastating weaknesses by stimulating the transmission of impulses between the various nerve and muscle groups. For some, it has relieved an annoying condition known as *tinnitus*, or ringing in the ear. This mineral also activates many enzymes that are necessary for proper absorption and utilization of many vitamins, including thiamine, biotin, and ascorbic acid (vitamin C). Manganese also assists the body's ability to produce sufficient "free radical scavengers" in parts of the body's cells. Thyroid glands depend upon manganese for normal functioning. Symptoms and disorders possibly caused by a manganese deficiency include convulsions, ringing in the ears, memory loss, inner ear imbalance, nerve disorders, and epilepsy. Insufficient levels of manganese can lead to a decreased ability to metabolize fats and carbohydrates. Some experts believe that a deficiency of manganese can be associated with connective tissue and cartilage problems. A high tissue content of calcium and phosphorus necessitates higher intake of manganese.

Toxic symptoms and effects of excess manganese were observed in industrial workers who were often exposed to its dust. Symptoms included weakness, impaired motor skills, psychological difficulties, and more. Manganese overload is often seen in alcoholics because it enhances the liver's manganese level, and is said to increase the liver's ability to absorb this trace mineral. The National Research Council sets the adequate dietary intake for manganese at 2.5 to 5 milligrams for adults. A few of the most abundant natural sources of manganese include egg yolks, nuts, seeds, green vegetables, whole grain cereals, kelp, and wheat germ.

MOLYBDENUM ∾ A hard, lustrous silver-white metallic element somewhat resembling lead, molybdenum was isolated in 1746 by the Swedish chemist P. J. Hjelm and named in 1781 by K. W. Scheele. In combination with other alloys, it is used for the manufacture of "points" such as for spark plugs. Molybdenum is another of the essential trace minerals found in most all plant and animal tissues. Among its many functions, it assists the body in metabolizing copper and is an integral part of certain enzymes, which help the body maintain a reserve of iron, all of which assists the body in oxidizing fats. The symptoms and effects of a molybdenum deficiency include male impotence, impaired ability to metabolize carbohydrates and fats, anemia, and a lowered growth rate. Excess molybdenum can cause diarrhea. Natural food sources include meats, cereals, legumes, and a few of the green leafy vegetables. The National Research Council recommends 150 to 500 micrograms of molybdenum daily.

NICKEL ∾ Nickel was discovered in or about 1754 by mineralogist A. F. Cronstedt. It is a hard, silver-white, malleable metallic element used in alloys, batteries, and plating, due to its resistance to oxidation. Nickel has been employed in the production of insecticides, nickel-plated faucets, cookware, safety pins, some jewelry, and electrical wiring.

Although it is determined to be an essential trace mineral, much of what is presently known regarding nickel is not flattering. Much more information has been discovered regarding problems associated with nickel overload than can be found regarding any benefits derived from this mineral's presence in the body. Food sources include margarine and shortenings, cocoa, and whole grains.

PHOSPHOROUS ∽ An essential mineral, phosphorous is the second-most abundant mineral found in the body. Present in every cell, it is involved in just about every chemical reaction. One part phosphorous to every 2.5 parts of calcium helps to maintain healthy bones. This required balance is upset in the presence of refined white sugar. Phosphorous is also beneficial in assisting the body to manufacture phospholipids, such as in lecithin, which aid the body in metabolizing and transporting fats. This mineral is also intricately involved in the synthesis of deoxyribonucleic acid (DNA), the coding which carries the messages for cell replication and life, and ribonucleic acid (RNA). Also, it has been utilized successfully to speed up the healing process of bone fractures and for the treatment of osteoporosis. Some research suggests that phosphorous may provide some protection against cancer. Deficiencies in phosphorous include metabolic disturbances, bone degeneration, changes in normal weight patterns, and appetite and/or memory impairment. There is no known toxicity associated with phosphorous intake. The RDA for phosphorous is from 800 to 1,600 micrograms, depending upon variable factors such as sex and age. If the body's calcium content is high, additional phosphorous can be taken to bring about a return of balance. Natural sources of phosphorous are meat, fish, whole grains, nuts, milk and dairy products, sesame seeds, and brewer's yeast.

PLATINUM ∾ This silvery, malleable, ductile metallic element is highly resistant to corrosion and electrochemical attack. Platinum is a chemical catalyst used for acid-proof containers, ignition fuses, and in the manufacture of jewelry. The human body has no known biological requirement for platinum.

POTASSIUM ∾ Named in 1807 by Sir Humphrey Davy, who first isolated it from potash, this soft, silvery-white, wax-like element oxidizes quickly when exposed to air and is found abundantly in nature in the form of its salts, which are used in fertilizers and glass. Potassium has long been utilized for the treatment of high blood pressure, as a sugar stabilizer for some types of diabetes, and for the relief of diarrhea in both infants and adults. Symptoms associated with a deficiency of potassium include nervous disorders, headaches, irregular heartbeat, insomnia, dry skin, muscle weakness, nausea, vomiting, and diarrhea. There is no known toxicity associated with potassium intake. The RDA has not been established. But many authorities recommend an intake of potassium should be between 2,000 to 2,500 milligrams. Natural sources for potassium include green leafy vegetables, millet seed, bananas, sunflower seeds, potatoes, garlic, raisins, yogurt, oranges, dairy products, and brewer's yeast.

RUTHENIUM ∽ This rare metallic element was named in 1828 by Estonian-Russian chemist G. W. Osann, who produced it in an impure form. It is an element of the platinum group and is very hard and brittle. Ruthenium has been used as a hardener in alloys of platinum and palladium. The human body has no known need for this mineral.

SCANDIUM ∽ Isolated from a source of ore, discovered and so named in 1879 by the Swedish chemist L. F. Nilson, scandium, a rare, silvery-white metallic element occurring with various elements of the rare-earth group, has been utilized for producing high-intensity light sources. There is no known requirement for the mineral scandium in the human body.

SELENIUM ∽ This gray, nonmetallic element of the sulfur group exists in many forms. Experiments on mice showed that selenium increases resistance to disease by increasing the number of antibodies that neutralize toxins. Dr. Julian E. Spallholz of the Veterans Administration Hospital in Long Beach, California, conducted this experiment. Others suggest that selenium may improve our energy levels, prevent or relieve arthritis, retard the aging process by attacking free radicals, and possibly prevent cataracts. Deficiencies of selenium may contribute to impaired growth, heart disease, cancer, premature aging, immune deficiency, hypertension, and infertility. Toxicity can occur at daily intakes of over 1,000 micrograms. High levels of selenium in soil have caused toxicity and some deaths in animals that grazed on the grains grown in such soil. Selenium can also contaminate water supplies located near irrigated soil, and toxicity has been reported through industrial inhalation.

Toxic symptoms include an increase in respiration rate, loss of hair, teeth and nails, loss of energy, paralysis, gastrointestinal distress, and even death. The National Research Council recommends a daily intake of 50 to 200 micrograms daily; however, doctors and other experts studying selenium suggest an intake of 100 to 400 micrograms. Natural sources of selenium include onions, tomatoes, broccoli, bran and wheat germ cereals, liver, eggs, whole grains, pumpkin seeds, soybeans, and brewer's yeast.

SILICON ∞ Modeled by T. Thomson, a Scottish chemist, in the early 1800s who noted its chemical similarities to boron, silicon is a nonmetallic element that is found in several forms and is always found in combination with other substances. Only oxygen, which combines with silicon to form silica, is more abundant in nature. Silicon is employed in the manufacture of transistors, solar cells, silicones, and ceramics. An essential mineral that is being studied today for its relationship to the health of the human body, silicon does play a critical role (along with several other minerals) in the health and maintenance of bone growth in the developmental stages of life. Also, a relationship exists to aging skin, as the degree of silicon in the body declines with maturity. Although deficiency is rare, symptoms include hearing problems, hair loss, softening of teeth and bones, and osteoporosis. Toxicity has occurred in miners who have been subjected to long hours of silicon inhalation without proper ventilation. Its effect upon the lungs has been likened to the effects of asbestos exposure. Natural sources of silicon include hard drinking water, plant fiber, and the herb horsetail.

SILICONE ∾ Silicones can be any of a group of polymerized, organic silicon compounds containing a basic formation of atoms that alternate and are usually accompanied by other organic groups that are attached to their molecular chains. They are characterized by their ability to resist changes that might be caused by water, heat, oil, and other compounds.

SILVER ∾ This extremely pliable and malleable white metallic element can be hugely brilliant when polished. One of the best conductors of heat and electricity, silver is considered to be a precious metal, and it is used for jewelry and silverware, among other things. The salt of silver has long been used in the manufacture of photographic film. While the dietary need for silver has not been established, small amounts in the body may be irrelevant. Silver is known to be an antagonist to selenium and vitamin E.

SODIUM ∾ A soft, silvery-white, alkaline metallic element with wax-like consistency, sodium was named in 1807 by Sir Humphrey Davy. Found in nature only in combined form, this element is extremely active chemically. An essential mineral in humans, it can be found in various places in the body, including the extracellular fluids, the vascular fluids within the blood vessels, arteries, veins, and capillaries, the intestinal fluids that surround the cells, and the bones. It functions with potassium to help the body balance its acid-alkali factor in the blood, and it also helps to regulate water balance in the body. Sodium with potassium is also essential in the processes of muscle contraction and expansion and nerve stimulation. The amount of sodium is required to be in perfect proportion with that of potassium, and since potassium is known to work in combination with other minerals to regulate the heartbeat (among other functions), it is easy to understand how large quantities of sodium can cause an imbalance of other essential minerals, increase the blood volume, and lead to hypertension. Furthermore, clinical studies indicate that low-sodium consumption is effective in preventing or relieving *toxemia* (bacterial poisoning), *edema* (swelling), *proteinuria* (albumin in the urine), and blurred vision. Since sodium is readily available in most foods, including table salt, a deficiency is virtually unseen today. Were this to occur, it could lead to a condition known as *acidosis*, digestive disorders, a weakening of the adrenal glands, and, of course, an imbalance of essential potassium. Our requirement for sodium is small and toxicity can result with an intake of over 14,000 milligrams. The best sources of organic sodium are sea salts such as kelp or dulse. Natural food sources include okra, celery, carrots, cucumbers, beets, asparagus, turnips, oatmeal, string beans, and strawberries.

STRONTIUM ✏ In 1808 strontium was named by Sir Humphrey Davy, who first isolated this pale-yellow metallic element. As with elements resembling calcium, its properties are found only in combination. Strontium compounds burn with a red flame and are used in fireworks. A deadly radioactive isotope of strontium (strontium 90) is present in the fallout of nuclear explosions. Experts disagree on the value of this mineral for the human body. For instance, studies reported by Dr. Stanley Skoryna, director of medical research at St. Mary's Hospital in Montreal,

Canada, indicated that strontium might be protective of certain energy-producing structures within the cell. However, Dr. David L. Watts, who operates a tissue mineral analysis laboratory in Texas, states, "Strontium has not been found to be necessary for normal biological functions and is not considered an essential element," and further states that "strontium is apparently antagonistic to calcium and can, therefore, interfere with normal calcium metabolism." Strontium is stable and one of the least toxic of the trace minerals. The National Research Council has not set a requirement standard for strontium, and there is no information available regarding natural sources.

SULFUR ✎ A pale yellow, nonmetallic element found in crystalline or amorphous form, sulfur burns with a blue flame and stifling odor and is used in vulcanizing rubber as well as in making matches, paper, gunpowder, insecticides, and sulfuric acid. Sulfur is also utilized in various forms in the manufacture of lotions, ointments, creams, and dusting powders. Considered an essential mineral to humans, sulfur is found in every cell of animals and plants. Body weight comprises up to 0.25% of sulfur. It is often referred to as nature's beauty mineral, because it keeps the hair glossy and smooth and the complexion clear and youthful. A sulfur deficiency can lead to disorders in the stomach, colon, throat, eyes, or liver, as well as dry brittle hair and nails, gas, bloating, and respiratory problems. Vegetarians could become deficient in sulfur if they refrain from eating eggs. There is no RDA for sulfur, as its requirement is presumably met when protein intake is sufficient. Toxicity could result with excessive intake.

Natural sources of sulfur are garlic, eggs, red cabbage, sprouts, figs, radishes, nuts, cauliflower, fish, and spinach. In supplemental form, sulfur can be found in garlic and the free-form amino acids, methionine, cystine, and cysteine.

TIN ∽ A soft, silvery white, crystalline metallic element, malleable at ordinary temperatures, capable of high polish and used as an alloy in tinfoil, solders, utensils, type metals, and magnets, tin is not an essential chemical to health. Much is yet to be discovered regarding tin and its presence in the human body.

TITANIUM ∽ Discovered in the late 1700s by British mineralogist William Gregor, this silvery-to-dark-gray, lustrous metallic element found in rutile and other minerals is used as a cleaning and detoxifying agent in molten steel and in the manufacture of aircraft, satellites, and chemical equipment. There is no essential need for titanium in the human body.

TUNGSTEN ∽ Named in 1755 by A. F. Cronstedt, tungsten was isolated from scheelite by J. J. De Elhuyar and F. De Elhuyar around 1783. In Swedish, "tung sten" translates as heavy stone. This hard, heavy, gray to white, metallic element found in wolframite, scheelite, and tungstite is employed in the manufacture of steel for high-speed tools, in electric contact points, and in lamp filaments. Tungsten is a nonessential mineral to the human body.

∽ ∽ ∽

VANADIUM ∞ Vanadium was named in 1831 by Berzelius for an element rediscovered by the Swedish chemist N. G. Sefstrom. First discovered by A. M. Del Rio in 1801 and best described as silvery white, this metallic element is alloyed with steel, due to its noted strength, and has been used in nuclear applications. Present in most of the body tissues and thought to have elements comparable to those of zinc, vanadium is believed to be essential to human health. In certain forms, such as *vanadyl sulfate*, it can be useful in lowering blood lipid levels (LDL cholesterol levels) and possibly in preventing atherosclerosis. Most of the vanadium is absorbed by the body or excreted in the urine; however, the liver is the most likely storage area for unused vanadium supplies. Excessive intake (over 300 micrograms per day) could prove to be toxic to some. There are no guidelines in place as determined by the National Research Council, but many experts or authorities suggest an estimated requirement of 100 to 300 micrograms per day will satisfy the body's requirements for vanadium. *Vanadyl sulfate* is the most utilized supplemental form of vanadium on the market today.

ZINC ∞ An essential mineral, zinc was first used in 1526 by Paracelsus, in the form of crystals in smelting. This bluish-white metallic element, usually found in combination, is used as a protective coating for iron, as a constituent in various alloys, as an electrode in electric batteries, and, in the form of salts, in medicine. In the body, zinc is beneficial in triggering the release of many essential enzymes and is also a component of insulin. It helps not only in the process of carbohydrate digestion but also in phosphorous metabolism.

Zinc is vital to proper growth and development and for optimum health of the prostate gland, the reproductive glands, and the immune system. Deficiencies lead to loss of appetite, loss of taste and smell (to some degree), impotence, stunted growth, impaired metabolism, and an inability to heal, among other problems. Zinc can become toxic in excessively high doses. Intakes over and above 660 milligrams (which were given to many elderly patients in the form of zinc sulfate) resulted in symptoms such as nausea and diarrhea. High dosages could also interfere with iron and copper utilization. The National Research Council recommends a dietary intake of 15 milligrams of zinc for adults, with an additional intake of 15 more milligrams during pregnancy or lactation. Other authorities suggest an intake of 50 to 100 milligrams. The average healthy intake may only yield about 8 to 11 milligrams of zinc a day. Natural sources of zinc include whole grains, pumpkin seeds (which are thought to be one of the highest natural sources for zinc and are utilized in the treatment of prostate problems), liver, eggs, meats, fish, poultry, mushrooms, soybeans, and brewer's yeast.

ZIRCONIUM ✑ Named by Sir Humphrey Davy in 1808, this hard, ductile, gray or black metallic element is found combined with zircon. Occurring in tetragonal yellow, brown, and red crystals, zircon is a mineral found in igneous and sedimentary rocks. It is used in alloys, ceramics, and in manufacturing some types of nuclear reactors. This is not a mineral that has been determined to be essential to the health of the human body.

VITAMINS

Vitamins are divided into two specific categories, water soluble and fat soluble. For the most part, excess water-soluble vitamins are usually excreted from the body via the urine or sweat glands. Fat-soluble nutrients, however, are absorbed into the body's fat, where they are stored for later use. Many of these fat-soluble vitamins have the potential for accumulating in various organs of the body, such as the liver, and becoming potential problems, or antagonists. So how does one determine the optimum dosage of a particular vitamin? The United States Recommended Daily Allotments are meant to be a guideline, but they cannot work best for everyone, because we are not all created equal when it comes to our needs and genetic requirements. Also, these recommendations vary or increase during times of stress, pregnancy, or disease. Megadoses of certain vitamins cannot always be stored properly in the body and can, as a result, actually cause harm. Nobel laureate Linus Pauling is well known for his theories on vitamin C and his belief in megadoses of specific vitamins.

To avoid vitamin imbalances, most practitioners prescribe individual vitamins, especially the B complex, in combination with other nutrients. Also, for the purpose of better assimilation and utilization of therapeutic doses of particular vitamins and nutrients, it is often recommended that daily doses be divided into two equal portions, preferably to be taken with meals.

Although vitamins are essential to life, the body might not show any deficiency symptoms for certain ones for as long as a year; however, we could not go as long as four months without vitamins B_{12} or D. The effect of such deficiencies would be reflected in liver functioning, such as its ability to store these vitamins and release them when needed. The liver has some 500 or so functions, and managing its vitamin banks is just one of them. The vitamins are listed here alphabetically.

AMYGDALIN ∽ See **Vitamin B_{17}.**

ASCORBIC ACID ∽ See **Vitamin C.**

BIOTIN ∽ Part of the B complex vitamins, biotin is water soluble. It is known as a *coenzyme*, meaning it is part of an enzyme system of constituents that work together to achieve a specific result. Some of those specific results include stimulation of the production of fatty acids, oxidation of fatty acids (as well as carbohydrates), and assistance to the body in its utilization of folic acid, protein, pantothenic acid, and vitamin B_{12}. Benefits of biotin supplementation include decreased hair loss and relief from certain types of skin diseases. Large or prolonged use of antibiotics can lead to a deficiency of this vitamin. Deficiency symptoms include dry skin, loss of appetite, lack of energy, muscular pain and/or insomnia. Another deficiency symptom is an increase in blood cholesterol levels or a lowered hemoglobin count. There are no known toxic effects associated with biotin. The National Research Council (NRC) sets recommendations for biotin to between 150 and 300 micrograms. Requirements for children range from 0.3 to 0.6 micrograms for infants to 1.4 micrograms for females from 11 to 14 years and 1.7 micrograms for males of the same age.

CHOLINE ∽ Another of the B complex vitamins, choline is also water soluble and is presently being studied for its effects on brain neurotransmission and memory. Available in most animal tissues (usually in combination with the other B vitamins), but mostly in combination with *lecithin* or *acetylcholine*, choline is required for the proper metabolism of fats, preventing them from becoming trapped in the liver. Often a slowed metabolism rate can be helped simply by adding a B complex vitamin to one's daily supplemental regime. In fact, I have personally witnessed how this can assist individuals in losing that extra 5 to 7 pounds after other methods have failed. A deficiency of choline has been linked to liver and kidney disorders. Choline is an essential nutrient required by all human cells. Its benefits include memory improvement, especially in some individuals diagnosed with Alzheimer's disease. It acts as a methyl donor, necessary for the transport of fat from the liver, thus having a profound positive effect on individuals with dangerous cholesterol levels and adding to the protection of the heart and arteries. Choline is available as a soluble salt or as a *phosphatidylcholine* such as *lecithin*. (Personally, I prefer lecithin, which is made from pure soy.) Choline's dosage recommended by most health practitioners stands at 1,000 milligrams; however, doses can range between 500 and 5,000 milligrams when used therapeutically for brief intervals. As with all of the other B vitamins, choline works best in the company of the other B complex vitamins. This is how we best avoid vitamin imbalances, unless prescribed in higher dosages therapeutically supervised by a qualified practitioner for a predetermined effect. No known adverse side effects are associated with supplementation of choline, when taken as directed. Natural sources for choline include whole grains, egg yolks, legumes, lecithin (phosphatidyl-choline), vegetables such as lettuce and cauliflower, and liver, soy, and, of course, brewer's yeast.

CYANOCOBALAMIN ∞ *See* **Vitamin B$_{12}$.**

FOLIC ACID (folacin) ∞ Folic acid is a water-soluble vitamin and part of the B complex of vitamins. Also often referred to as B$_9$, folic acid works to assist the cells in their growth and regeneration process. It is also essential for the formation of red blood cells and aids in the metabolism of protein. In conjunction with B$_{12}$ and vitamin C, folic acid helps the body to produce *heme*, the iron protein that is part of the hemoglobin, a requirement for the formation of red blood cells. Folic acid is the carbon-carrier. Recent studies show that *homocysteine* is a factor in the progression of both *ather-osclerosis* and osteoporosis, and it has been shown to in-crease the risk of heart attack. Clinical trials indicate that supplementation of folic acid actually lowers *homocysteine* levels in most individuals who fall within the high area of predetermined risk zone. Folic acid deficiency symptoms include fatigue, shortness of breath, heart palpitations, nausea, weakness, irritability, headache, and anemia. Certain medications such as estrogens, alcohol, barbiturates, cer-tain drugs used in chemotherapy (especially *methotrexate*), as well as the effects of excessive stress can interfere with the absorption of folic acid. There is no known toxicity associ-ated with supplementation of folic acid; however, excessive intake can often disguise itself as a vitamin B$_{12}$ deficiency.

As with all of the other B vitamins, folic acid should always be taken in combination with all of the other B vitamins. The RDA for folic acid has been established at 180 micrograms for an adult female and at 200 micrograms for adult males. Pregnant and lactating women may require slightly higher amounts. The requirements for children vary from 25 micrograms (for babies 6 months and older) to 75 or 100 micrograms for children 6 to 10 years of age. These requirements often do not compensate for the lack of certain specific nutrients caused by genetic weaknesses or diagnosed disorders. Such considerations, in my professional opinion, could preclude the necessity of taking some prescribed medications and offer a better and more reliable guide, perhaps, than an "across the board" single recommendation meant to address all individuals. Folic acid can be found in high content in such foods as asparagus spears, beef liver, broccoli, collards, mushrooms, oatmeal, peanut butter, red beans, and wheat germ. Folic acid is destroyed at high temperatures and by exposure to light, and it becomes ineffective if left at room temperatures for extended periods.

INOSITOL ⁓ Now accepted by most as a part of the vitamin B complex vitamins, *inositol* is often found with choline and works intimately with biotin. Inositol is also associated with vitamin B_6, PABA, folic acid, and pantothenic acid in helping those vitamins thoroughly perform many vital responsibilities associated with the nerve, brain, and muscle functions. Inositol is an element abundant in lecithin, which helps to remove fats from the liver and prevent liver and artery diseases, because it assists the body in excreting the buildup of stagnated fats and bile. Also, inositol accounts for that part of fiber compound known as *phytic acid*. Finally, it is proven to be of value to some diabetics who suffer neuropathy (a long-term complication of that disease). Deficiency symptoms include high blood lipid levels (high cholesterol levels) and artery and coronary diseases. There are no known side effects associated with the use of inositol. The Recommended Daily Allotment (RDA) for inositol has not been established. However, inositol has been used therapeutically (without any known side effects) in the treatment of depression or panic disorder (12 grams daily for short term only), for the treatment of diabetes, at doses of 1,000 to 2,000 milligrams, and for removing toxins from the liver, 100 to 500 milligrams daily. Natural sources of inositol include citrus fruits, seeds and legumes, whole grains, oatmeal, nuts, molasses, and brewer's yeast. It can be purchased commercially as inositol monophosphate.

LAETRILE ∾ See **Vitamin B$_{17}$.**

LINOLEIC AND LINOLENIC FATTY ACIDS ∾ See **Vitamin F.**

MENADIONE ∾ See **Vitamin K.**

NIACIN ∾ See **Vitamin B$_3$.**

NITRILOSIDES ∾ See **Vitamin B$_{17}$.**

OROTIC ACID ∾ See **Vitamin B$_{13}$.**

∾ ∾ ∾

PANAGAMIC ACID ∽ *See* **Vitamin B$_{15}$.**

PANTOTHENIC ACID ∽ *See* **Vitamin B$_5$.**

PARA-AMINOBENZOIC ACID (PABA) ∽ This is a naturally occurring drug used in lotions and creams to prevent sunburn. It was formerly administered by mouth to treat certain infections now treated with antibiotics. As part of the B complex vitamins, it is recognized for stimulating growth, as well as treatment of skin disorders such as eczema and lupus erythematosus. Symptoms of a deficiency include eczema, fatigue, loss of libido, anemia, and *vitiligo* (a skin disease that causes the skin to lose its natural pigmentation). Finally, it is said to help individuals with premature graying of the hair to find a return of color. High doses may cause nausea, vomiting, itching, and rashes. PABA is always available in the tissues and is synthesized by friendly bacteria within the intestines, provided the environment is conducive. PABA is a vital constituent of the B complex and is available in combination with folic acid. Para-aminobenzoic acid is, however, not a vitamin in itself, therefore, no RDA requirements have been formerly established at this writing. High potencies of PABA are available only through prescription, because many researchers believe that continuous ingestion of high doses of PABA can be toxic!

PYRIDOXINE ∽ *See* **Vitamin B$_6$.**

RIBOFLAVIN ∽ *See* **Vitamin B$_2$.**

THIAMINE ∽ *See* **Vitamin B$_1$.**

TOCOPHEROL ∽ *See* **Vitamin E.**

VITAMIN A ∞ This fat-soluble vitamin is necessary for the repair and growth of body tissues. It works in protecting the mucous membranes of the mouth, nose, throat, and lungs from possible infections, and it is also necessary for the health and integrity of the linings of the digestive tract, including the kidneys and bladder. Individuals who spend long hours watching television or working with computers might require more vitamin A, because the mucous membrane tissues of the eyes can become irritated through constant glare, poor lighting, or eyestrain, such as can occur when continually working on computers. There is some protection provided for computer workers in the attachment of a filtering screen known as an "eye saver."

Vitamin A is absorbed via the upper intestinal tract, where the fat-splitting enzymes and bile juices change it into a usable nutrient. The majority of vitamin A is stored in the liver, and tiny amounts are distributed to the fat tissues in the kidneys, lungs, and retinas of the eyes. Under unusually stressful conditions, the body will quickly use up its reserve supply. Diets containing less than the recommendations for fat necessary for proper body functioning could result in inadequate amounts of bile reaching the intestine, causing a vitamin A depletion through the feces.

The benefits of vitamin A include protection from infections (including the eye, vagina, ovaries, and intestines), conversion of male and female hormones, and improvement of such skin conditions as acne, ulcers, and eczema. Toxicity symptoms include hair loss, itchy, dry, flaky skin, blurred vision, nausea, and enlargement of the liver and spleen. In order to avoid toxicity, use vitamin A in its *beta-carotene* form, as it is better tolerated by the body, more efficiently stored, and time-released as the body requires it. Beta-carotene is also a powerful antioxidant, (*see also* Chapter IV, Supplements). The RDA for vitamin A is 1,500 to 4,000 international units (IUs) for children and 4,000 to 5,000 IUs for adults. Food sources include carrots (which contain 11,000 IUs), or calves' liver (which contain 74,000 IUs).

VITAMIN B COMPLEX ∽ Among the B vitamins are B_1 *thiamine*, B_2 *riboflavin*, B_3 *niacin*, B_5 *pantothenic acid*, *pantetheine*, B_6 *pyridoxine, folic acid, choline, biotin, inositol,* and B_{12} *cyanocobalamin.* The B complex vitamins are water soluble, making them non-toxic, because any unusable amount will be excreted by the body through normal processes.

Today's modern American diet, over-abundant in refined white sugars and processed foods, is not only devoid of the B complex vitamins, which are lost in these processes, but is in itself antagonistic, meaning it "opposes" or is counterproductive to a desired natural process, contributing to a depletion of one's reserve supply of many essential nutrients. Also, because of the presence of food additives and preservatives (inorganic chemicals), the natural process of metabolism and utilization of nutrients becomes a more difficult task. Many of these additives and preservatives are said to adhere to the walls of the intestines and colon, contributing to evacuation problems, accumulative toxins within the colon, and constipation, the number one cause of colon cancers today. The FDA states that the average individual at the age of 45 or 50 has an accumulated fecal matter attachment (to the intestine and colon) of anywhere from 5 to 20 pounds!

Care needs to be taken in supplementing with the B vitamins. Doses of each may vary depending upon age, sex, caloric intake, and/or the desired effect. Except under supervision, individual B vitamins should never be supplemented alone, as this could trigger an imbalance among the other B complex vitamins. Normally a 100 milligram dose of the B complex vitamins is recommended for adults, and lesser amounts for children. Proper nutrition will eliminate any possibility of a deficiency, other than those attributable to genetic tendencies. Discussion of each of the vitamins in the B complex follows in alphabetical order with the other vitamins.

Sources for the B complex vitamins as a whole include alfalfa sprouts, asparagus, green vegetables (such as broccoli, brussels sprouts, cabbage, and others), black-eyed peas, celery, corn, eggplant, green beans, whole grains, eggs, fish, liver, and meats. While many foods contain most of the B complex vitamins, they may not contain them all. Brewer's yeast is a food source that does contain all of the B complex vitamins, as well as other essential elements.

VITAMIN B₁ (thiamine) ∞ Thiamine functions in part by aiding enzyme production in the citric acid cycle, also known as the Krebs cycle, a complex cycle of enzyme-related catalyzed reactions, occurring within the cells of all living animals. Acetate, in the presence of oxygen, is broken down to produce energy in the form of an electron transport chain. This action also is the final step in the process of oxidation of carbohydrates, fats, and proteins. In combination with *niacin*, thiamine has been used to treat multiple sclerosis.

Thiamine deficiency can result from inadequate dietary intake or from malabsorption. One particular disease resulting from a vitamin B₁ deficiency is *beriberi*. Beriberi takes two forms—wet berberi, a condition characterized by an accumulation of tissue fluid (edema), and dry berberi, extreme emaciation. There is nervous degeneration in both forms of the disease, and death from heart failure is often the outcome. Beriberi is not common in the United States, although alcoholics may develop cardiac (wet) berberi with congestive heart failure, neuropathy, and/or cerebral disturbances. Other deficiency symptoms of vitamin B₁ include nerve problems, heart muscle weakness, and edema. A diet high in refined white flours, refined white sugars, and junk foods can cause a depletion of the B vitamins. First signs of this may be becoming more quickly fatigued. Other symptoms noted have included cardiovascular problems, including heart muscle weakness, and a decrease in eye-hand coordination. There is no known toxicity associated with thiamine; however, overdosing of any one of the B vitamins may cause an imbalance of one or more of the other B vitamins. The National Research Council (NRC) requirements are 0.5 milligrams of thiamine per 1000 calories consumed, for individuals of all ages. Vitamin B₁ is a water-soluble vitamin that can be obtained through foods such as whole grains, meats, and legumes.

VITAMIN B₂ (riboflavin) ∞ As is the case with all of the B vitamins, riboflavin is a water-soluble vitamin. Riboflavin works in conjunction with thiamine in aiding enzymes in the citric acid cycle, also known as the Krebs cycle.

Riboflavin's presence in the B complex helps to correct symptoms associated with malabsorption as well as certain types of eye and skin disorders. Deficiency symptoms include inflammation of the skin and eyes. Although taking huge doses of supplementation on a continued basis may stimulate extraordinary urinary losses, along with the depletion of the other B complex vitamins, under normal conditions, there is no known toxicity associated with riboflavin. The NRC sets the daily requirements for riboflavin at 1.6 milligrams for the adult male and 1.2 milligrams for the adult female. The daily requirement for children is as follows: 1–3 years, 0.5 milligrams; 4–6 years, 1.0 milligram; 7–10 years, 1.2 milligrams; 11–14 years, 1.3 milligrams for females, and 1.5 milligrams for males; 15–18 years, 1.8 milligrams. Body size and caloric intake may require adjustments in the requirements of this B vitamin. Riboflavin is usually abundant in green vegetables, organ meats, eggs, and dairy products.

VITAMIN B$_3$ (niacin) ∽ Niacin is yet another of the water-soluble B complex vitamins. More stable than either thiamine or riboflavin, niacin is more resistant to heat, light, air, acids, and alkalies. It is manufactured synthetically in 3 forms—*niacinamide, nicotinic acid,* and *nicotinamide.* Niacin is another of the B vitamins that works together with thiamine and riboflavin in completing the Krebs cycle. This vitamin is also effective in improving circulation and reducing cholesterol levels in the blood. Deficiencies of niacin have been associated with skin and nervous system disorders. No known toxicity results when taken in recommended doses, although large amounts have been reported to cause liver damage. Certain forms of niacin, however, can cause a temporary flushing of the skin, and, in doses of over 100 milligrams, have been associated with depression in some individuals; therefore, while niacin is essential, it is best taken only at the recommended dosage as set by the NRC, 6.6 milligrams per 1,000 calories consumed. For men, this amounts to approximately 16 milligrams, for women 13 milligrams, and 9 to 12 milligrams for children, depending upon age. The best known source for niacin is brewer's yeast, although it is available in small amounts in meats and grains.

VITAMIN B₅ (pantothenic acid) ∞ A water-soluble vitamin and part of the B complex, pantothenic acid assists those enzymes involved in the metabolism of fats and carbohydrates. Vitamin B₅ is also beneficial to those with rheumatoid arthritis, those in need of adrenal support, and those individuals who have been diagnosed with high cholesterol and triglyceride levels. While the symptoms of a deficiency of pantothenic acid are rare, loss of coordination could become noticeable. No significant side effects have been associated with either pantothenic acid or pantetheine (another readily available source of B₅). The RDA for supplementation of vitamin B₅ in children 11 years of age or older is 4 to 7 milligrams. The adult requirement for this vitamin is a minimum of 10 milligrams daily for adults. Brewer's yeast, a good source for vitamin B complex, is very easily assimilated by the body. Vitamin B₅ can also be found in organ meats, eggs, and liver.

VITAMIN B₆ (pyridoxine) ∾ Another B complex vitamin, vitamin B₆ is water soluble and plays a vital role in assisting enzymes that are critical for the proper completion of a process referred to as *catabolism*, whereby living tissue is changed into energy and waste. Pyridoxine is often beneficial for certain types of depression. It also works in stimulating multiplication of all cells, in helping to keep the immune system functioning properly, and has been utilized therapeutically (only under professional supervision and in specific available forms) for asthma, autism, cardiovascular disease, carpal tunnel syndrome, diabetes, epilepsy, nausea, osteoporosis, and premenstrual syndrome. Symptoms of a deficiency involve irritability, anemia, or convulsions. Excessive or prolonged supplementation of B₆, without the company of all of the other B vitamins, can harm nerve functioning and cause an imbalance in many of the other B vitamins. Excessive intake of vitamin B₆ has also been reported to cause symptoms related to such conditions as nerve damage, depression, and more; therefore, it is strongly recommended that anyone contemplating high doses of pyridoxine, seek the advice of a qualified practitioner. The RDA is 0.9 to 1.6 milligrams (per 1,000 kilocalories of dietary intake) for children, 2.0 milligrams for women, and 2.2 milligrams for men. Therapeutic doses as recommended by professional practitioners may vary greatly, depending upon variable factors such as medical diagnosis, dietary intake, symptoms, and age. Antagonists to B₆ include food colorings (FD and C yellow no. 5); some medications, such as *hydralazine, dopamine, penicillamine,* and oral contraceptives; alcohol; and a diet which includes an extremely high intake of protein. Vitamin B$_6$ can be found in brewer's yeast, sunflower seeds, wheat germ, soybeans, walnuts, soybean flour, lentils, lima beans, navy beans, brown rice, garbanzos, bananas, avocados, whole wheat flour, chestnuts, kale, spinach, turnips, sweet peppers, potatoes, raisins, sweet potatoes, and cauliflower.

VITAMIN B$_{12}$ (cyanocobalamin) ∽ One of the most extraordinary of all the vitamins, vitamin B$_{12}$ is the only one that also contains essential mineral nutrients. It is a water-soluble vitamin that contains cobalt, an element said to be associated with longevity. Also unlike the others, it cannot be manufactured synthetically (that is, produced through chemical synthesis, rather than from natural and available sources). Instead, it is grown in bacteria and molds. Folic acid and vitamin C are necessary for the completion of the metabolism of vitamin B$_{12}$ in order for cells to develop and mature. Because it is more involved in the production of blood than in any other process, without it, pernicious anemia could be the result. Deficiency symptoms include a weakness or soreness in the arms or legs or difficulty in walking or speaking patterns. Nervousness and neuritis (a disease of the peripheral nerves showing the pathological changes of inflammation) are often common complaints, as is memory loss. Under the care of a qualified practitioner, vitamin B$_{12}$ has been used successfully to treat such diseases as pernicious anemia, sprue, fatigue, asthma, and hepatitis. There have been no known cases of vitamin B$_{12}$ toxicity. The RDA requirements for vitamin B$_{12}$ are 3 micrograms for adults and 4 micrograms for pregnant and lactating women. Infants require a daily intake of 3 micrograms, while growing children require 1 to 2 micrograms. As we begin to age, the absorption of B$_{12}$ is decreased. The vegetarian is often deficient in B$_{12}$ and folic acid. Sources for vitamin B$_{12}$ include kidney, fish, muscle meats, dairy products, and brewer's yeast.

VITAMIN B$_{13}$ (orotic acid) ∞ Necessary for the biosynthesis of nucleic acid, this vitamin is essential for the regenerative processes which occur within the body's cells. It has been valued mostly for its benefit in the treatment of multiple sclerosis. Deficiency symptoms are not known at this time, but a lack could eventually lead to liver disorders and cell degeneration, including premature aging. RDA requirements are not known. Sources for vitamin B$_{13}$, or orotic acid, include whey (that portion of milk that remains after much of the fat has been removed) and soured milk.

VITAMIN B$_{15}$ (pangamic acid) ∞ Another water-soluble vitamin, vitamin B$_{15}$ is most valued for its unique ability to restore oxygen to living tissue, especially in the condition known as *hypoxia*. It works in conjunction with some of the other B vitamins for better metabolism functioning, enhanced circulatory performance, and maintenance of proper health of the nervous and glandular systems. Deficiency symptoms may not be immediately noticeable, but prolonged unavailability of pangamic acid may contribute to heart and/or glandular diseases. Used therapeutically in medical practice in Russia and other countries, but not in the United States, vitamin B$_{15}$ has no known toxicity. Although the RDA for this vitamin has not been established, the usual therapeutic dose is 100 milligrams a day—50 milligrams in the morning before breakfast and 50 milligrams at night before retiring. Sources of vitamin B$_{15}$ include whole grains, seeds, nuts, and whole brown rice.

VITAMIN B$_{17}$ (nitrilosides, amygdalin) ∞ Vitamin B$_{17}$ is also known as *laetrile* when used in medical dosage form. Among its known benefits is its ability to prevent and control free radicals, as proposed by its discoverer Ernst T. Krebs, Jr., who discovered the Krebs cycle (*see* Vitamin B$_1$). Therapeutic doses are determined by doctors using vitamin B$_{17}$ in the treatment of cancer. Prolonged deficiencies are thought by many to lead to a diminished resistance to malignancies. The NRC has not yet accepted this vitamin as official or essential to human nutrition, and therefore no standards have been set. Natural sources include vegetables; most whole seeds from fruits such as apples, apricots, peaches, plums, raspberries, cranberries, blackberries, blueberries; many beans and grains, including mung beans, limas, garbanzos, millet flour, buckwheat, and flaxseed.

VITAMIN C (ascorbic acid) ∞ Thought by most experts to be non-toxic, this water-soluble vitamin is found in many fruits and vegetables. It is said to be able to boost the immune system's action, flush toxins from the body, and help to prevent certain vitamin deficiencies. Vitamin C (ascorbic acid in its natural form) is often used in conjunction with other vitamins, herbs, and supplements, as in a formula. It has synergistic ability, that is, it assists the body in more quickly metabolizing other nutrients and making them available for use. Vitamin C is involved in many bodily functions, but its primary responsibility is thought to be that of maintaining collagen, a protein necessary for the building of connective tissue. It also assists in the production of red blood cells and in the prevention of hemorrhaging. Vitamin C fights bacterial infections. Because it reduces the effects on the body caused by many allergy-producing substances, it is often used in preventing the common cold. Also, vitamin C maintains a significant relationship with other essential nutrients in the body, including the metabolism of the amino acids *phenylalanine* and *tyrosine*. Vitamin C also changes folic acid into its active form of *folinic acid*. It is said that this vitamin also plays an important role in the metabolism of an essential mineral, calcium. Excess vitamin C carried to the bladder may help in the prevention of bladder cancer.

Vitamin C has proven itself to be beneficial in aiding the process of digestion, lowering the incidence of infections, removing toxins from within the body, and maintaining and protecting the immune system. In addition, it may be useful in helping to reduce blood cholesterol levels of patients with arteriosclerosis. Some say that it can often help prevent colds and infections or lessen their severity. Several universities and practitioners are evaluating vitamin C's benefit to painful and swollen joints. A deficiency can lead to a disease known as scurvy, which results in an inadequate production of collagen, an extracellular substance that binds the cells of the teeth, bones, and capillaries. This disease was common among sailors and others deprived of fresh fruits and vegetables for extended periods of time. Rare today, scurvy could very well develop in individuals on restricted diets (such as alcoholics), or babies who are taken from being breast-fed and placed on cow's milk, without being provided with adequate amounts of vitamin C. Saturation of vitamin C in the human body is said to be about 5,000 milligrams, and 30 milligrams are utilized by the adrenal glands, 200 milligrams are taken up in the extracellular fluids, and the remainder is distributed in varying percentages throughout the body's cells.

Vitamin C is the least stable of the vitamins and very sensitive to oxygen. It is, however, fairly stable in acid solutions. Its potency can be lost through exposure to light, heat, and air, which stimulate the activity of oxidative enzymes. The body's ability to absorb vitamin C is greatly hindered by cigarette smoking, stress, illness, or extended periods of antibiotic or cortisone use. Other causes of an inability to absorb vitamin C have been attributed to exposure to toxic chemicals such as DDT or the fumes of petroleum. Aspirin and some painkillers are also vitamin C inhibitors, and sulfa drugs increase the urinary output of vitamin C by as much as three times the average amount. Baking soda creates an alkaline medium that destroys Vitamin C. Drinking excessive quantities of water will also add to the excretion of the body's vitamin C content. Finally, vitamin C is destroyed through the using of cooking utensils that contain copper. For detailed information on supplementation, *see* Chapter IV, Supplements, Vitamin C.

VITAMIN D ∽ Vitamin D, also called *ergosterol*, *viosterol*, or *calciferol*, is essential for bone formation and the absorption of calcium. A particular form of vitamin D (found in maitake mushrooms) has also been recognized as beneficial for the treatment of certain types of leukemia. Because of the three forms of this vitamin and the effects of each, vitamin D has been called both a vitamin and a hormone. A deficiency of vitamin D is usually associated with senior citizens or shut-ins. This is due to the fact that our bodies are able to produce vitamin D through the stimulation of the sun. With the sun's stimulating effect, a precursor to vitamin D called D_7-dehydrocholesterol is converted into vitamin D_3. It is then carried to the body's main filtering organ, the liver. Once there, the liver, with the aid of enzymes, converts D_3, or cholecalciferol, into another, more potent type of the D vitamin, known as 25-dihydroxycholecalciferol. The last transition is made on this second type of D_3 by enzyme participation from the kidneys. When this occurs, 25-dihydroxycholecalciferol (or 1,25-OH2D3) is the end result.

While once it was thought that calcium alone was required for the formation of healthy bones and the treatment of osteoporosis, we now know that to retard the progression of many bone diseases, or in some cases, to actually reverse the degeneration process, a lot more Vitamin D is needed. Calcium, glucosamine sulfate, vitamin C, vitamin D, boron, and silicon are part of the new and revised protocol. Without vitamin D, rickets and skeletal deformity could result. The deficiency may show early symptoms, such as profuse sweating, irritability, and restlessness, while a chronic deficiency may induce multiple bone malformations due to their softening, bow-legs, and knock-knees. Eventually, chronic deficiencies could lead to pain in the lower back and legs, with difficulty walking or climbing stairs. On the negative side, prolonged supplementation of vitamin D has caused elevated levels of calcium and phosphorus in the blood and excessive excretion from the urine, which can lead to calcification of the soft tissues of the walls of the blood vessels and kidney tubules (better known as "*hypercalcemia*"). When given in large doses to individuals diagnosed with rheumatoid arthritis, vitamin D encouraged the development of calcium deposits in the arteries, a development that can contribute to kidney disorders or high blood pressure.

Since most of our needs for vitamin D are fairly minimal (unless one is diagnosed with a disease which requires higher amounts in balanced proportions and in combination with other essential vitamins and minerals), we need not be too concerned regarding our body's supply of this vitamin, and most experts agree that vitamin D should not be supplemented unless recommended or prescribed by a practitioner. In general, it is thought that children and pregnant women require 400 IUs daily, but differences of opinion exist in the case of the elderly. Some believe that the elderly (who are in cessation of bone growth) require less vitamin D and that 200 IUs are sufficient, yet other experts believe that, because a good portion of the elderly are less active outdoors, their need for vitamin D does not decrease. Sources of vitamin D include dairy products and fish liver oil. Some vitamin D can be found in butter, egg yolks, and dark green leafy vegetables. *Ergocalciferol* (D_3) is the vitamin D type most often used when adding nutritional supplements to cow's milk. Vitamin D is also absorbed through the skin, an effect made possible through activation of a form of cholesterol, a direct result of the sun's ultraviolet rays. Hence, its name, *the sunshine vitamin*.

VITAMIN E (tocopherols) ∾ Vitamin E is a fat-soluble vitamin. While there are several forms of *tocopherol* (*alpha*, *beta*, *delta*, *epsilon*, *eta*, *gamma*, and *zeta*), *alpha tocopherol* is thought to be the strongest form of vitamin E and the form that is most beneficial to humans. The many benefits associated with vitamin E supplementation could fill pages. Briefly, vitamin E is an antioxidant; it can attack and literally overcome free radicals in the body, which are often the result of oxidation and can cause cancer. Also, it is beneficial and recommended for disorders involving the circulatory, skeletal, respiratory, muscular, glandular, or neurological systems. In addition, it is vital to the health of the skin, and it assists the body in protecting cell membranes from being catabolized (broken down). Vitamin E is proven to reduce the side effects from some painkillers such as *codeine*, *morphine*, and *aminopyrine*. The *Journal of the American Medical Association* has reported that vitamin E is useful for diminishing the side effects (such as hot flashes) associated with menopause. Deficiencies in vitamin E are not easily recognized, but edema and/or skin lesions in infants and muscle weakness, including intermittent claudication in adults, have been noticed. Other deficiencies of vitamin E include ruptured blood cells, muscular wasting, reproductive disorders, coronary artery diseases such as strokes and/or heart attacks, as well as a weakened immune system. Chronic or prolonged deficiencies could result in blood vessel fragility and a reduction or shrinkage of collagen (connective tissue). While for the most part, vitamin E is considered non-toxic, caution should be taken for supplementation in individuals who suffer from hypertension, as vitamin E in large doses can raise the blood pressure.

In individuals who are being treated for heart problems (such as rheumatic conditions), large or excessive doses (more than 400 IUs) could cause further destruction of that organ, or lead to a fatality. Recommendations for vitamin E include 12 IUs for females and 15 for males over the age of 11. However, for pregnant or lactating women, 15 to 18 IUs is the suggested supplement. Because vitamin E is an antioxidant, which helps protect the body from cell changes caused by free radicals, it is sometimes recommended at higher dosages (such as 400 IUs) to combat the effects of toxins (or carcinogens) that bombard us each day in the form of chemicals, herbicides, pesticides, food additives, animal fat, hydrogenated oils, and more. Natural sources of vitamin E include seeds, nuts, whole grains, polyunsaturated oils, asparagus, avocados, berries, and green leafy vegetables.

VITAMIN F (linoleic and linolenic fatty acids) ∽ The benefits of vitamin F are numerous. It helps to protect the arteries and prevent arteriosclerosis, evacuate the buildup of excess cholesterol (therefore, protecting the heart), promote calcium absorption, and protect the body from damage caused by exposure to radiation. Deficiencies of the essential fatty acids could effect the skin (such as in eczema), the hair, the adrenal glands, the reproductive glands, and the prostate gland. The National Research Council states that our daily fat intake should always include the essential fatty acids (the good fats) as part of our caloric intake and defines 1% of caloric intake of vitamin F as adequate. Natural sources include unprocessed and unrefined oils such as flax oil and sunflower oil. See more about essential fatty acids earlier in this chapter.

VITAMIN K (menadione) ∽ Vitamin K is essential for the manufacture of *prothrombin*, a substance that stimulates coagulation. Without prothrombin, hemorrhaging could occur and lead to death. There are 3 specific types of vitamin K. K_1, referred to as *phylloquinone*, is the one thought to be most beneficial. K_2 is derived from plant sources and contains the fat-soluble nutrient chlorophyll. Without vitamin K_2, derived from a fat-soluble plant source, the essential normal process known as photosynthesis (as it pertains to humans) cannot be accomplished. (See essential fatty acids for more information earlier in this chapter.) K_3, or *menaquinone*, is a synthetic form of the natural vitamin K.

Vitamin K is absorbed easily in the digestive process, and the synthetic form is employed for use in individuals who lack the enzymes necessary for the assimilation and utilization of the natural form. Deficiency symptoms include frequent nose bleeds and loss of vitality, leading to internal hemorrhages. Its absence has also been linked to premature aging. The benefits of vitamin K are rapid healing from cuts or wounds, protection from premature aging, and providing a healthy supply of energy. Excessive amounts of the synthetic form can have serious side effects, and, because of that fact, this version is only available through prescription. The RDA allotment for vitamin K is recommended in micrograms. Male children over 11 years of age require 45 micrograms, ages 15 to 18 require 65 micrograms, and 19 to 25 years and older need 70 to 80 micrograms. Young females over the age of 10 are thought to need about the same as is recommended for boys of equal age; however, females 15 to 18 years require about 10 micrograms less than the opposite sex. Sources of vitamin K include egg yolks, cow's milk, liver, oats, asparagus, and whole wheat. Vitamin K can also be found in such food sources as many of the green colored plants, including spinach, kale, mustard greens, asparagus, and broccoli.

VITAMIN T ∽ Vitamin T, or the "sesame seed factor," as it is also known, is very much in the discovery stage, as biochemists are still studying and reviewing its need and benefit for the human body. Vitamin T is useful for reversing anemia caused by nutritional deficiencies and in treating hemophilia by stimulating the production of blood platelets. Vitamin T is non-toxic, and, at the present time, there are no known established recommendations as set by the RDA. Sources include sesame seeds or tahini, egg yolks, and a few vegetable oils.

VITAMIN U ∽ Considered more as a co-vitamin factor, vitamin U is found most abundantly in cabbage. The benefits of cabbage are many, but due to its known healing and protective powers to the colon, practitioners who favor natural therapies often suggest that either raw cabbage, cabbage juice, or homemade sauerkraut (without all the sodium) be included at meals twice weekly. There are no deficiency symptoms or toxicities associated with cabbage. It works instead as a preventative and as a healing property.

∽ ∽ ∽

DIGESTIVE ENZYMES
Nature's Catalysts for the Utilization of Nutrients

According to *Bantam's Medical Dictionary*, an enzyme is a protein that, in small amounts, speeds up the rate of a biological reaction without itself being used up in the reaction, acting as a catalyst. An enzyme acts by binding with the substance involved in the reaction, the substrate, and converting it into another substance. The names of enzymes usually end in "ase," and they are named according to the substrate upon which they act (as in lactase) or the type of reaction they catalyze (as in hydrolase). Each enzyme is relatively specific in the type of reaction it catalyzes, and each requires certain conditions for optimum activity. These include the correct temperature and pH, the presence of certain coenzymes, and the absence of specific inhibitors. Enzymes comprise two basic elements; one is a protein molecule, the other a coenzyme. A coenzyme is a non-protein organic compound that, in the presence of an enzyme, plays an essential role in the reaction catalyzed by the enzyme. Coenzymes usually include the B vitamins in their molecular structure. Produced within living cells, enzymes may act either within the cell, as in cellular respiration, or outside it, as in digestion. Enzymes are necessary for proper functioning of such processes as digestion, absorption, metabolism, and necessary bodily changes. Enzymes known as proteases can stimulate certain immune system cells.

Although enzymes are stable, heat or certain chemicals easily inactivate them. Enzymes are so essential for the normal functioning and development of the body that failure in the production or activity of a single one may result in metabolic disorders. Some of these are inherited, and others are the result of the normal process of aging, insufficient dietary intake of raw vegetables, which stimulate production of enzymes, and excessive intake of over-cooked foods. Digestive enzymes belong to a group known as hydrolyzing enzymes or hydrolyses. The following are the digestive enzymes that we should be most concerned with:

LIPASE ∽ This enzyme is responsible for the breakdown of fats. Fats are not water soluble, therefore they must go through an intensive process of being broken down into smaller molecules that can be digested. Lipase accomplishes this task through stimulating the release of bile, which works in the small intestine to metabolize fat.

PROTEASES ∽ These enzymes catalyze the hydrolysis of proteins, changing them into a compound which makes them ready for the next process of digestion, breaking them down further into compounds known as peptides, and finally into amino acids. The three main proteases are *trypsin*, *pepsin*, and *peptides*, and all work at various levels to accomplish a single goal, digestion.

PTYALIN ∽ This is the main enzyme in saliva, and it works optimally in a neutral to somewhat acid pH. This enzyme, which performs the initial task of breaking down food, is initiated from impulse centers in the brain by stimulation of the senses. These stimuli are sent to the brain, which in turn sends a message to the salivary glands to secrete saliva.

Chapter III: Dietary Sources

A balanced diet should supply most if not all of our daily nutritional needs. The following is an alphabetical list of food sources with brief descriptions of each:

ALMONDS ❧ Abundant in potassium, magnesium, phosphorus, protein, and laetrile, these are the fruit seeds of sweet almond trees. Because laetrile acts as an anti-cancer agent, cancer clinics around the world often include 10 raw almonds per day in their recommended protocol. Almond oil or almond butter is highly nutritious. One ounce of almonds contains 15 grams of fat, of which only one is saturated.

APPLES ❧ "An apple a day keeps the doctor away" is to be taken more seriously than the old nursery rhyme. The American Journal of Clinical Nutrition reports that apples have been shown to decrease the time it took to have a bowel movement by increasing stool weight, which in turn increased the number of trips to the bathroom during a 24-hour period. Canadian scientists (the December 1978 issue of Applied & Environmental Microbiology) have demonstrated that fresh apple juice or fresh applesauce could knock out stomach flu and polio viruses. The Annals of Nutritional Metabolism stated that apple fiber extracts containing a high level of pectins decreased the level of cholesterol in hamsters. Eating apples can prevent cholesterol-induced gallstones from forming. The main constituent of apples is water, but they also contain vitamins A and C.

APRICOTS ∞ Originating in Central Asia, apricots have been utilized for everything from skin softeners to wrinkle removers, with some medicines being developed from the kernels. They are high in vitamin C. Due to its beta-carotene content, the juice of the apricot has a definite advantage over some other fruit juices.

ARROWROOT ∞ Arrowroot is a wonderful grain for diabetics, as it is very low in carbohydrates, and therefore, does not greatly effect the blood sugar level. This grain has been around for centuries but is virtually unknown in the United States.

ASPARAGUS ∞ Rich in vitamins A, B complex, C, potassium, manganese, and iron, asparagus also boasts high amounts of histones, and folic and nucleic acids, which stimulate the immune function, and it is an anti-carcinogenic. Asparagus contains an important element, the amino acid *asparagine*, which stimulates the kidneys, resulting in increased urine, an occurrence that normally follows the intake of large quantities of fluids.

AVOCADO ∞ High in protein, the avocado's oil contains vitamins A, D, and E, 14 minerals, and is a rich source for copper and iron. Avocados also contain phosphorus, magnesium, calcium, sodium, and manganese. Avocados' potassium content is in balance with its sodium, making it a good choice for potassium (more potassium can be found in an avocado than in a banana.) Avocados help to stabilize the blood sugar levels of hypoglycemics.

BANANAS ∾ Low in fat, bananas are a good source of vitamin C and potassium.

BEANS, DRY ∾ Beans contain a good portion of daily fiber and help to satisfy hunger. Beans are similar to grapes in that their nutritional values can be rated by their color. The darker the color, the more benefit can be derived. High in fiber, beans help to prevent some forms of cancer, especially colon cancer. Diabetics can also benefit by incorporating small servings of beans into their diets, as they help to slow down the escalation of blood sugar levels. Eating beans too frequently, however, can contribute to a buildup of gas within the intestines. Pre-soaking does help to eliminate that problem. More information on individual types of beans follows.

BEANS, ADZUKI ∾ Native to China and Japan, these tiny red beans cook quickly due to their size. Those on a calorie-restricted diet will benefit from incorporating these beans into their casseroles or soups, because, ounce for ounce, these beans hold less calories than most. They are believed to be highly nutritious.

BEANS, GARBANZO (chickpeas) ∾ These beans are often used in salads and can make a great dip. They are abundant in minerals, especially calcium, potassium, and iron. They are also a good source of vitamin A. My favorite way of enjoying these beans is to rinse them under running water, empty into a shallow baking dish, and add a tablespoon or two of olive oil and cayenne pepper to taste. Bake at 325 degrees for about 20 to 25 minutes and enjoy.

BEANS, GREEN ∽ These beans are a good source of hemi-cellulose, a constituent that is found in various fruits and vegetables, such as beets, whole grains, brussels sprouts, cabbage, corn, peppers, apples, and pears. They are said to be beneficial for weight loss and prevention of colon cancer.

BEANS, KIDNEY ∽ The kidney bean is high in protein and is said to surpass all others in its high fiber content. They do contain other nutrients as well, but they are utilized most for their fiber and protein content.

BEANS, LENTIL ∽ Actually, the lentil belongs to the pea family. Lentils are abundant in nutrients such as protein, calcium, magnesium, phosphorus, sulfur, and vitamin A. Because they are without a distinctive taste of their own, any favored ingredient, such as garlic, will blend well.

BEANS, LIMA ∽ As with most of the other types of beans, limas have many nutrients to offer, including minerals, fiber, and vitamins. They work well as a casserole (after presoaking overnight, and parboiling) with tomatoes, onions and Canadian bacon.

BEANS, PINTO ∽ Pinto beans contain as much as 15 grams of protein per cup, and that is about 6 grams per 100 calories. The high fiber content in beans make them very satisfying and an excellent choice to add to a soup or as a side dish.

BEANS, SOY ∽ High in the mineral zinc and vitamin E, soybeans also contain linoleic acid. Soy is a good source of protein. Learn more about soy's versatility under such headings as tofu, soy milk, and soy flour.

BEETS ～ Rich in calcium, phosphorus, sodium and potassium, iron, and magnesium, beets also contain high amounts of amino acids. They aid lymphatic functioning, gallbladder and liver problems, digestion, and building red blood cells in anemia cases. During cooking, minerals become concentrated, while vitamins A, B complex, and C are lost.

BERRIES ～ The berry family includes boysenberries, blackberries, blueberries, dewberries, currants, elderberries, huckleberries, loganberries, gooseberries, raspberries, and strawberries. Considered by herbalists to be a great detoxifier, berries are a good source of fiber, and many varieties contain anti-cancer compounds as well as vitamin C and potassium. Blueberries contain an enzyme known as bilberry, which has been proven in clinical studies to be a potent liver protector and to increase the circulation to the small capillaries behind the eyes, making them of use in the treatment of many eye diseases. Blackberries, golden raspberries, and strawberries are all good sources of fiber and can often be utilized as an ingredient substitute for processed sugars in baking. Some individuals have shown allergic responses to strawberries. Blueberries and blackberries are rich in vitamin A, while blueberries also contain some iron, potassium, magnesium, and silicon.

BRAZIL NUTS ～ Brazil nuts are fruits of a tall evergreen tree that grows wild in the Amazon. Resembling the coconut, the creamy, meaty kernels of brazil nuts are high in calcium, phosphorus, and thiamine (part of the B vitamin complex).

BREWER'S YEAST ∞ *See* Chapter IV, Supplements.

BROCCOLI ∞ A vegetable that is abundant in vitamins A, C, and E, as well as selenium (a powerful antioxidant), broccoli is said to lower one's risk of cancer. This is due, in part, to some of its many beneficial constituents, such as sulforaphane, indoles, dithiolthiones, and glucosinolates, including its abundant source of carotenoids (Vitamin A). Broccoli has no fat and is very low in calories.

BUCKWHEAT ∞ This whole grain is often sold under the names of *kasha*, or *kasza*, and has a very distinct flavor. Buckwheat does contain some gluten, but not nearly as much as wheat. Often used for making pancakes, it contains fiber and several essential minerals. This grain is known for its ability to quiet hunger for many hours, leading to its ability to aid in weight loss. Because buckwheat is organic (alive), it must be refrigerated.

BUTTER ∞ A dairy product made from churned cream, butter is very high in monounsaturated fats, as well as saturated fats. Butter contains the vitamins A and D. Because butter has an extremely high concentration of fat, especially the saturated fats, it can contribute to atherosclerosis (or artery blockages), and be a leading factor associated with heart attacks or strokes. Olive oil is a healthier substitute, and for a special treat, try ghee. Ghee is a type of butter that has had most of its saturated fat removed. This is done through a process of first heating the butter until it is completely melted, and then allowing it to sit covered until it has cooled down sufficiently. At this point, the fat rises to the surface and can be easily skimmed, or poured away. What is left is a very tasty butter, with half of the saturated fat content. Tahini or almond butters are other healthy choices. While they, too, contain a fair share of fat, it is less saturated, as it is derived from vegetable sources, rather than animal. Another pleasant change from butter, and a way to cut the amount of saturated fat in one's diet, might be to replace the butter used on top of a baked potato with mashed avocado. Those who suffer from hypoglycemia will appreciate this tip, as avocados help to stabilize the blood sugar.

CABBAGE ❧ One of the magnificent 12 cruciferous vegetables, raw cabbage contains an abundance of chlorophyll, a substance which is beneficial in preventing anemia. Cabbage also contains vitamin A and sulfur. It is low in calories, protein, and carbohydrates, but provides a good bit of fiber. Cabbage is thought to be an excellent blood purifier, and is healing to ulcers while stimulating to the immune system. Many experts believe that eating cabbage once a week (including the juice left after cooking) can cut the risk of colon cancer by more than 50%. If you are a nursing mother, and your baby has a problem with colic, be aware that brussels sprouts, cabbage, and cauliflower, as well as yeast breads and dairy products may be the culprits. Some of their constituents, which cause allergic reactions in some individuals, are passed to the child through breast milk. The "magnificent 12" cruciferous vegetables include broccoli, brussels sprouts, cabbage, cauliflower, horseradish, kale, kohlrabi, mustard greens, radishes, rutabagas, turnips, and watercress.

CAULIFLOWER ∽ Another of the cruciferous vegetables, cauliflower is a good source of sulfur as well as smaller quantities of carotenes and chlorophyll. Cauliflower also contains a fair share of vitamin C. It has no fat and is low in carbohydrate content. This low-calorie food is often recommended for diabetics.

CARROTS ∽ Raw carrots are a good source of beta-carotene (a precursor of vitamin A), vitamin B complex, C, D, E, K, iron, calcium, phosphorus, sodium, potassium, magnesium, manganese, sulfur, and copper. They are a powerful antioxidant and help to prevent heart disease, reduce the risk of cancer, help to lower LDL cholesterol, improve eyesight, and prevent eye and mucous membrane infections.

CASHEWS ∽ Grown in Africa, India, and South America, cashews are an excellent source of iron and folacin (part of the B complex vitamin), as well as a good source of vitamin A, potassium, and magnesium. While other nuts may contain more fat than the cashew, the proportion of saturated fat is relatively high (compared to almonds or walnuts, for example). Therefore, it is not a good choice of nut for frequent snacking. One ounce of dry roasted cashews contains about 13 grams of fat, broken down into 3 grams of saturated fat, 7.8 grams of monounsaturated fat, and 2.2 grams of polyunsaturated fat.

CELERY ∽ A good source of both vitamins and minerals, celery contains vitamins A, B, C, and E, along with sulfur, calcium, and potassium. One of the best natural diuretics available, celery helps the body to excrete carbon dioxide.

A couple of stalks a day is beneficial to the adrenal glands and can aid in weight loss. Celery offers some fiber and practically no calories at all.

CHEESE ∽ Cheese is the end result of separating the whey from the milk curd. The curd is that part which becomes somewhat solid, while the whey is that part which is more liquid. With proper aging, the desired texture and flavor will occur. Cheese usually contains protein, calcium, phosphorus, riboflavin, and vitamin A. Various types of processed cheeses offer many tastes from tart to sweet, and from soft to firm. All, unfortunately, are high in saturated fat, sodium, and calories. In moderation, cheese can compliment most meals.

CHERRIES ∽ Fat-free sweet treats, sour cherries are lower in calories and higher in vitamin C and beta-carotene than are sweet cherries. In fact, they contain enough beta-carotene to provide 26% of the RDA as compared to their cousin the sweet cherry, which contains only 12% of the RDA. In 1992, *Natural Health* reported in their March/April edition that ellagic acid, which is present in cherries and strawberries, counteracts both human and naturally occurring cancer causing agents.

CHESTNUTS ∽ The U.S. supply of chestnuts is imported from Europe, where most of these nuts are grown. They are composed mainly of carbohydrates and are low in fat and calories, but are starchy when first picked. After a few days of curing, much of the starch is converted to sugar, and these big soft nuts develop a gentle sweetness. One ounce of roasted chestnuts contains only one gram of unsaturated fat.

CHILI PEPPERS ✑ Although chili peppers are an irritant to the stomach, they signal the bronchial cells to pour out fluids, making the lung and throat secretions less thick and sticky. Because of this, they may be useful for individuals suffering from asthma and hypersensitive airways. The *capsaicin* in chili peppers (which is the substance that makes them hot) has been found to reduce the swelling of the tracheal and bronchial cells caused by cigarette smoke and other irritants. Capsaicin is also a painkiller, causing a reduction in nerve cells of substance P, which relays pain sensations to the central nervous system. This food also aids in preventing as well as dissolving blood clots.

COCONUTS ✑ Another fruit seed, like the almond, coconuts grow on the tropical coconut palm tree and come to the United States from Central America and Puerto Rico. While most of us usually never get to see the actual fruit, we are familiar with the egg shaped, hairy shell it grows in. Though coconuts are relatively high in fiber, there is not much in mineral benefit available. They are also fairly high in calories, including some saturated fat.

COLLARDS ✑ An abundant source for vitamins A and C, this vegetable is a good source for chlorophyll and is a relative of kale.

CORN ✑ Rich in vitamins A, B, C, potassium, iron, zinc, and magnesium, corn is also high in fiber. It is considered a brain food and is good for the nervous system and may also be good for those needing to gain weight. Research from the University of Nebraska reports the quality of protein in corn is better than nutritionists once believed. Yellow corn is best, containing the most nutrients. For some, corn is an allergy-producing food and should be avoided by those who have digestive problems or are on a weight-loss diet.

CRANBERRIES ∽ An excellent source of fiber, cranberries have also been utilized for centuries because of their healing effect on kidney and bladder disorders. In 1984, *The Journal of Urology* revealed that cranberry juice is an excellent inhibitor of bacterial adherence in the urinary tract as well as a reliever of bladder infections in females. It has long been considered an intestinal antiseptic, possibly attributable, in part, to its high vitamin C content.

CUCUMBERS ∽ A whole cucumber has only about 55 calories, no fat, and its fair share of vitamin A and iron. Cucumbers stimulate urination and are said to be a good blood cleanser.

CURRANTS ∽ Black currants are an excellent source of potassium and fiber, yet low in calories. They are often substituted for raisins, but can be more difficult to come by.

DATES ∽ Dates are a good source of fiber and contain many other nutrients, including some of the essential amino acids. They can be found in a powdery consistency and make a healthy (yet somewhat more expensive) alternative as a sweetener. These tasty fruits are, however, high in calories also. Moderation is the key.

∽ ∽ ∽

EGGS ⁓ The debate over eggs and how they fit into the diet is not over yet! Eggs are high in protein and contain vitamins B_1, B_2, B_5, B_{12}, biotin, choline, E, and K. Eggs are also a good source of manganese, sulfur, iron, phosphorus, choline, and lecithin. Choline and lecithin, found in the yolk, are what helps to keep the cholesterol moving in the bloodstream. While eggs contain cholesterol and some authorities recommend avoiding them, it is a fact that the body does require some cholesterol in order to function properly. Some may point to the egg as a potential threat for E-coli bacteria, but if the egg is thoroughly cooked, that possibility is removed. People who like eggs might be able to purchase eggs obtained from free-running hens, fed organic feed, and without any added drugs or hormones.

FIGS ⁓ Figs are rich in calcium and iron, and may be among the oldest cultivated fruits. Like dates, figs are noted for their soft texture and sweetness. They are also a good source of vitamins, minerals, and fiber. Once harvested, they last for only about one week, which is why much of our supply is dried. Figs are also known for their ability to destroy intestinal parasites. They contain 17% more calcium than skim milk, but 629% more calories!

FILBERTS and HAZELNUTS ⁓ Many people believe that the filbert and the hazelnut are interchangeable; however, the filbert is a small European tree, while the hazel, a wild shrub, is its close American relative. Turkey supplies the United States with filberts. They are high in potassium, sulfur, and folacin, as well as some calcium and magnesium. One ounce of hazelnuts contains 18 grams of fat, one of the saturated type.

FLAXSEED ∞ Rich in the essential fatty acids, omega 3, and omega 6, flaxseed is the highest, purest form of the essential fatty acids. One tablespoon a day will supply a sufficient amount, which need not constitute more than one or two percent of the daily caloric intake. Either flaxseed or flaxseed oil can usually be found at health food stores; however, because flaxseed is subject to very rapidly turning rancid, it is advisable that it only be ground at the time of ingestion. For that reason, flaxseed oil is often more preferable, as this cold-pressed oil can be refrigerated. *Flaxseed oil should never be heated!*

FRUITS ∞ The USDA (United States Dietary Guidelines) recommends we include four fruits each day. In fact, there probably is no known fruit that does not provide needed vitamins and nutrients. One of the first benefits gained from a diet rich in fresh fruits is oxygen. Oxygen is the very essence of life, and provides us with energy. For more details on specific fruits, see their separate listings.

GARLIC (*Allium sativum*) ∞ Native to Asia, garlic grows wild in Italy and France. It is a member of the lily family, closely related to onions, leeks, scallions, and chives. Literature of the great ancient kingdoms of Babylon and Medo-Persia, Greece and that of Rome began the praises of garlic that continue to the present. The great naturalist and writer Pliny recommended garlic for intestinal disorders, dog and snake bites, asthma, tuberculosis, convulsions, tumors, and scorpion stings. For centuries, garlic has been used in China and Japan for treating high blood pressure. In 1858, Louis Pasteur noted the mild antibacterial action of garlic.

Near the turn of the century, reports confirmed that garlic was showing remarkable effects in the treatment of tuberculosis. During both world wars, garlic was successfully used as an antiseptic and disinfectant to prevent infection and gangrene in wounds. Dr. Albert Schweitzer, while working as a medical missionary in Africa, used garlic to treat cholera, typhus, and amoebic dysentery with positive results. In Russia today, garlic is used extensively in treating various infections; in fact, it has earned the name of "Russian penicillin." In the 1940s, Dr. Arthur Stoll, a chemist working in Switzerland, was able to extract an oil from garlic that he named "allin." He also discovered an enzyme in the garlic, to which he gave the name "aminase." Aminase was found to change the allin to allicin when the garlic was cut or crushed. It is the allicin that is responsible for the garlic odor as well as for its antibacterial properties. Allicin is such a powerful oxidizer and disinfectant that, even when diluted with water by 1/80,000 or even as much as 1/120,000, it is still able to kill the germs that cause cholera and typhoid fever. Garlic contains more allicin than do any of its close relatives in the onion family. Garlic juice has been found to be active against fungi and yeasts as well as bacteria, including antibiodic-resistant varieties. Because garlic contains germanium and selenium, garlic was studied in the 1950s and 1960s as a possible cure for cancer, with encouraging results. While the majority of results are positive, keep in mind that if garlic is used in excess for long periods of time, it could cause imbalances. For therapeutic doses, seek out a professional in the field of natural healing, rather than attempting a long-term self-treatment plan. Some studies have shown that when laboratory rats were fed high doses of garlic, some developed anemia, experienced weight loss, and suffered stunted growth.

GRAINS ∽ Grains have been a staple of the diet since the dawn of humans. A country's economic status greatly influences the amount of grain that is normally eaten. As each country develops economically, its people are better able to purchase more grain for animal feed and thus consume more meat. Less stable countries, however, are dependent upon their grain crops for feeding their families. Traditionally, richer countries have always consumed more meat. It is not commonly known that the amount of land necessary for raising enough cattle to feed one individual is enough to grow wheat that can satisfy the hunger of fifteen.

The complex carbohydrates provide more than half of all the calories contained in grains, exactly the amount that should comprise about two-thirds or more of the calories we consume each day. Grains are also rich in both soluble and insoluble fiber. Soluble fiber helps to lower LDL blood cholesterol levels, while insoluble fiber prevents constipation and an accumulation of colon toxins, making it a protector against some forms of cancer. People living in areas where unrefined whole grains make up a significant part of the diet are alleged to have a lower incidence of intestinal cancer, diverticulosis, and hemorrhoids as compared to individuals living in industrialized countries. Whole grains and grain products also offer abundant amounts of B vitamins (riboflavin, thiamine, and niacin), vitamin E, iron, zinc, calcium, selenium, and magnesium. Most grains are also a good source of protein, without the fat. The diet in many un-der-developed countries consists of as much as 50% whole grain, a good source of their protein intake. Europeans and North Americans put less significance upon whole grains in their diets. In fact, Americans consume about 15% of whole grains as part of our protein intake.

Each grain is unique. Barley, corn, millet, oats, rice, rye, triticale, and wheat belong to the grass family, Gramineae. Each of the various other grains belongs to other botanical families. A kernel is an edible seed composed of three parts—the endosperm, the bran, and the embryo. Rice, oats, and barley are grains enveloped by a thin, inedible tissue covering, referred to as the hull, which must be removed prior to processing grain. Inside of each kernel are those nutrients essential for the embryo's growth and maturity. When the kernel has completed sufficient growth and has reached maturity, it is able to take root and obtain nour-ishment from outside sources. The bran is the outer layer of the kernel. It does not make up as much as half of the grain itself, yet it is able to develop an interesting profile made up of a few very distinct layers, such as the aleurone. These bran layers are rich in nutrients, especially many of the B complex vitamins. Whole grains almost always contain the bran and are usually packed with fiber.

In the United States, most breads, pasta, rolls, and flours are made from refined white flour, which amounts to over 60 or 70% of grain consumed in this country. This type of flour yields products with a longer shelf life. While these products may provide some nutrients, they definitely lack much of the dietary fiber and some of the nutrients found in the whole grains. In the process of refining whole grains, most nutrients are lost. The fiber-rich bran is stripped away during milling. White flour, when refined, for example, can lose up to 80% of the vitamins and minerals present in the whole-wheat kernel, and retains only 25% of the fiber. The FDA requires that a few of these nutrients (in their synthetic forms) be added back again, but usually at very minimal percentages and without regard to imbalances. Zinc and copper, for instance, as well as many of the other trace minerals are removed in the processing of whole grains. From a nutritional profile, therefore, high-quality whole grains are still superior to the refined ones.

Whole grains often need to be kept refrigerated in order to lengthen their life span, and this expense is another reason why the local supermarkets do not usually stock them. However, your local health food stores do carry them in their refrigerated sections. If you are lucky, you may be able to find local, freshly made whole grain bread in your health food store as well. These breads will also keep in the freezer indefinitely. A good way to introduce them to your family is to substitute the refined white flour with a mixture of 75% unbleached organic white flour and 25% flour such as spelt, kamut, and soy. Whole grains that have been around for centuries, but are virtually unknown to most of the population in the United States include spelt, kamut, millet, amaranth, barley, soy, and arrowroot. Arrowroot is a wonderful grain for diabetics, as it is very low in carbohydrates, and therefore, does not greatly effect the blood sugar level.

Certain grains provide additional specific benefits. Millet flour, for example, is a good source of B vitamins, and ounce for ounce is a better source of potassium. Oat flour is also very nutritious, and, when added to the diet, it may help lower high blood lipid levels. Buckwheat contains very little gluten (a major component of wheat grain), which is a source of allergens for many susceptible individuals. When buckwheat is added to the diet, it keeps hunger at bay and can help to reduce food cravings in those individuals who are prone to overeat. Quinoa is often referred to as a super-grain due to its high levels of potassium and riboflavin, but it is also a good source of magnesium, zinc, copper, manganese, and folacin. Rye is a good source of protein and may be similar in appearance to wheat, except for its bluish-gray color.

GRAPEFRUIT ∾ This is a good source of vitamin C, but only the red varieties contain beta-carotene. It also is a good source of pectin (known to lower cholesterol and now being studied for its ability to fight cancer). Grapefruit is low in calories and fat and is high in fiber.

GRAPES ∾ For centuries grapes have been utilized by herbalists for their ability to fight off the effect of toxins and their cleansing effect upon all of the body's tissues and glands. This fruit reduces edema and is a good source of vitamin C.

GRAPESEED ∾ A distinct plant species, grapeseed was first discovered in the eighteenth century in Barbados as a mutant from a Southeast Asian citrus fruit, the pomelo or shaddock, which had been brought to Barbados by a trader in the 1600s. Grapeseed extract has been shown in experiments to assist in the healing of candida (yeast infection) and other types of infections. It works quickly and safely and has also been shown to be effective in fighting viruses and ridding the body of parasites. This is truly a good natural alternative to pharmaceutical antibiotics. It should never be taken orally in a straight solution as it could burn the mouth, throat, or stomach. Capsules are also available. Side effects for some individuals include mild stomach irritation or flatulence.

HAZELNUTS ∽ *See* **Filberts**.

KALE ∽ Kale is one of the magnificent 12 cruciferous vegetables. It is high in vitamins A, C, E, magnesium, calcium, niacin, iron, riboflavin, sulfur, sodium, chlorophyll, potassium, phosphorus, and selenium.

KELP ∽ This is one of various large brown seaweeds and one of the richest sources for nutrients that can be found. It contains essential minerals, more than 40 trace minerals, iodine (which stimulates the thyroid gland), protein, essential fatty acids, calcium, and chlorophyll. Kelp also makes an excellent substitute for traditional table salt, sodium chloride. It also works as a chelator in that it attaches itself to and removes excess heavy metal within the body. (High levels of aluminum, lead, mercury, or others can become toxic.) Other varieties of sea vegetables include kombo, wakame, hijuki, and nori. Eating too much kelp is never a good idea; excess ingestion of it can trigger a headache. *See also* Chapter IV, Supplements.

KIWI FRUITS ∽ *See* **Fruits.**

LEMONS ∽ *See* **Fruits.**

LIMES ∽ *See* **Fruits**.

MACADAMIA NUTS ∽ Named after their discoverer, Dr. John Macadam, the Australian who reputedly discovered that they were edible and delicious, these nuts are high in fat and calories, yet they do contain significant amounts of iron, magnesium, and thiamine. One ounce of roasted macadamia nuts has a whopping 22 grams of fat, 2 of which are saturated.

MEAT ∽ Meat is a valuable source of protein, providing all of the essential amino acids, and many required minerals. Each meal should comprise 30% protein. Lean beef is a source of many versatile dishes. Because of what we have learned in recent years regarding the potential adverse effects associated with a diet high in saturated fats (high blood lipid levels, coronary artery disease), many have turned away from red meats. Lean meats that are eaten in moderation and baked or broiled (without a lot of oil) are still a vital part of one's daily nutritional requirements. Once again, moderation is the key. Three ounces of red meat (about the size of the palm of one's hand) is all that is required at each meal in order to obtain sufficient protein. It is increasingly easier for consumers to find meats from livestock raised on chemically-free grains.

Individuals who opt for a vegetarian lifestyle are often at a higher risk for developing anemia, insufficient protein intake, and possibly several other related conditions. It is important for people to understand the necessity of attaining a sufficient daily protein intake. Cooked rice and corn in combination provide a good source for protein, as does corn and beans, or soy.

MELONS ∽ *See* **Fruits.**

MILK ∞ Most of U.S. milk is from cows, though goat's milk is sold as well. Milk is high in calcium and vitamin D. As with any other type of food, there is the good and the bad: Most childhood allergies are traced to milk. Children who demonstrate hyper-sensitivity to dairy products may also be allergic to soy milk. Healthy alternatives with less potential for triggering allergic responses are almond drinks or rice milk.

MILLET ∞ This grain is one that is usually well tolerated by individuals who have an allergic response to wheat. Similar to other grains, millet is lacking the amino *lysine*; however, eaten with beans or legumes, its protein content is enhanced. This is a grain that contains no gluten and is, therefore, agreeable to those who suffer from a gluten intolerance. Millet is a pale grain almost without a distinct taste of its own. It can be combined with other grains. After simmering, millet resembles rice and can be added to hamburger or become an ingredient in meatloaf. It can also make a good dessert that could easily compete with the traditional rice pudding. It, too, is high in many minerals, especially folacin and phosphorus, and it contains other B vitamins, with only one gram of unsaturated fat per serving. Because it lacks gluten, this grain cannot be utilized for making breads, unless combined with another grain that offers some gluten.

MOLASSES ∞ There are a few varieties of molasses; however, most have gone through a lengthy processing, which often removes many of the natural nutrients. *Blackstrap molasses* is the heartiest source of molasses and preferable by health experts. Many of the B vitamins are abundant. It also offers many minerals, such as phosphorus, iron, and magnesium.

It has more calcium than milk and also contains vitamin E. Just one tablespoon of blackstrap molasses provides 3 milligrams of iron and more than 100 milligrams of calcium. Truly, here is a healthy substitute for refined white sugar. Blackstrap molasses can be added to cereals or homemade baked beans. It has been used successfully in the treatment of anemia, eczema, colitis, and many types of nervous disorders.

MUSHROOMS ∞ Mushrooms are a good source of protein and zinc. There are many varieties of mushrooms, such as porcini, chanterelle, cepe, morel, portabella, shiitake, maitake, and, of course, button mushrooms. The only mushroom which does not have any known benefit is the button mushroom. In fact, it contains substances known as *hydrazides*, which are cancer-causing. These hydrazides are rendered harmless through cooking. Portabellas lend themselves wonderfully to a pesto sauce for a versatile way of getting sufficient protein without red meat. Mushrooms have been used for decades by herbalists for enhancing the immune system, and the shiitakes provide an effective antiviral substance known as *lentinan*.

MUSTARD ∞ A condiment prepared from the oil of the mustard seed, mustard usually has a strong and pungent taste. Prepared or processed mustard is combined with vinegar, water, salt, turmeric, and paprika. It is very low in calories, and has no fat.

OATS and OATMEAL ∞ A whole grain that is loaded with fiber, about 5 grams per serving, no fat, and only about 110 calories per serving, oats also provide a good source of iron, copper, folacin, manganese, and vitamin E. Fiber helps to keep the bowels healthy, prevent colon cancer, and lower cholesterol. There are two types of fiber, soluble and insoluble, and oats contain both forms. Oats also contain many of the B vitamins, and those with a gluten intolerance will be happy to learn that oats have no gluten.

OILS ∞ Extra virgin or virgin olive oil has been found to be the healthiest of the cooking oils. Hydrogenated oils are often seen on the ingredient labels of processed foods, in everything from baked goods to frozen french fries. A diet high in hydrogenated oils has been associated with high free radical activity. In the hydrogenating process, an unsaturated (or good fat) is exposed to extremely high temperatures which break down the molecular structure of the oil, turning it into a saturated fat (or bad oil), creating an oil which hardens fat and destroys the essential fatty acids while releasing free radicals. Again, this process is commonly used to prolong a product's shelf life, but it can also produce a cancer-causing effect. See Essential Fatty Acids in Chapter II for information regarding the best source of cold-pressed oil.

OKRA ∞ This is a tall annual of the mallow family that is cultivated for its green pods. A soothing agent food, okra protects the internal membranes, relieving irritations of the linings of the digestive tract. It reduces inflammation and is a good source of fiber.

∞ ∞ ∞

ONIONS ∾ Offering vitamin C, onions, like their relative garlic, are a good source of organic sulfur. According to Tufts University, onions have the ability to increase HDL (the good cholesterol) while reducing the LDL (or bad cholesterol) up to as much as 25 or 30%. Another benefit of the onion is its ability to reduce the levels of fibrinogen (clot-forming substance in the blood), which may be beneficial in reducing the risk of stroke, especially for those who suffer hypertension. In fact, it might be said that onions offer more benefit therapeutically than nutritionally. The sulfur contained in onions helps to antagonize heavy metals and assists in their release. Onions are an excellent source of selenium; and they may also protect against stomach cancer.

PAPAYA ∾ This fruit has long been hailed for its ability to render harmless toxic acids in the body, including uric acid. A buildup of uric acid has been associated with a disorder known as gout. Papaya aids in the digestion process and contains vitamins A and C.

PARSLEY ∾ In small amounts, parsley provides an extra rich source of vitamins A, B_1, B complex, C, as well as potassium, manganese, phosphorus, calcium, and iron. An aid to all bodily functions, it assists digestion, sweetens breath, works like a diuretic, stimulates the adrenal and thyroid glands, and is an excellent cleanser.

PEACHES ∾ Easily digestible, peaches contain fiber, potassium, and beta-carotene (vitamin A).

PEANUTS ∾ Peanuts are a complete protein, but they have the highest fat content of all nuts. In addition, they may be contaminated by a compound called aflatoxin, a known carcinogen. One ounce of peanuts roasted in oil has 14 grams of fat, 2 of which are saturated. Two tablespoons of peanut butter has 16 grams of fat, of which 3 grams are saturated. Peanut oil contains 18% saturated fat, 48% monounsaturated fat, and 34% polyunsaturated fat.

PEAS ∞ A link between the consumption of peas and low rates of acute appendicitis has been shown in studies conducted in Wales and in England. This vegetable is believed to possess a chemical that suppresses organisms that cause infection in the appendicular wall. Peas are high in anti-fertility agents, work to lower blood pressure, and help to control blood sugar. They contain high amounts of fiber, carotene, and vitamin C. They are fat free, and a good cancer fighter. Peas also help lower LDL cholesterol.

PECANS ∞ The seeds of a species of hickory native to North America, pecans grow wild from Illinois to the Gulf of Mexico. Pecans are high in potassium and vitamin A and are abundant in essential fatty acids, or good unsaturated fats. However, they are the lowest in protein of any of the nuts. Dried pecans (1 oz. serving) contain 19.0 grams of fat, with only 1.5 grams being saturated (or of the bad type of fat), 12.0 grams monounsaturated fat, and 4.5 grams of polyunsaturated fat. There are no resources available for pecan oil, as it is not commonly used.

PEPPERS (red and green bell) ∞ High in vitamin C, peppers are good for all types of illnesses. Since cancer and most degenerative diseases thrive in an acid environment, sweet peppers are a good substitute for citrus fruits.

PINE NUTS ∽ These chewy, sweet nuts are delicious when added to salads or fruit combinations. The European version is richer in protein and lower in fat than the American version, but the American pine nuts offer more vitamins and minerals.

PISTACHIOS ∽ Often tiny holes made by a worm can been seen in these nuts. Pistachios have beige shells and green kernels, which are rich in thiamine, iron, and phosphorus. One ounce of pistachios has 14 grams of fat, 2 saturated.

POTATOES ∽ This versatile and filling vegetable is has become a favorite American side dish. Baked, boiled, mashed, fried, or in a casserole, potatoes have become a regular staple in many households. Just a little over 100 calories, a baked potato has one gram of fat and no cholesterol. It provides vitamins B_6 and C, and several minerals such as potassium, manganese, iron, copper, and magnesium. The potato is somewhat high in carbohydrate content, however, a factor diabetics need to be aware of.

POULTRY ∽ A good source of protein, poultry is another versatile food. It can be ground up for casserole dishes or baked whole. It can be breaded and fried or grilled. Chicken and turkey are low in fat, especially the white meat with the skin removed. Chicken breast that has been cooked without the skin contains about 160 calories and 4 grams of fat (only one gram of saturated fat.) Poultry contains several vitamins and minerals.

Perhaps the biggest concern regarding poultry today is safe handling. Only purchase poultry that has an expiration date that meets with your personal comfort zone. Be sure to cook it before that expiration date, and keep your refrigerator thermostat set cool enough to maintain healthy standards. Prewarm your oven, or skillet, before unwrapping poultry. Never allow poultry to thaw on the kitchen counter; instead, defrost with the microwave or in the refrigerator prior to using. With a reliable disinfectant, thoroughly clean your sink, counter top, hands, and anything else that may have come in contact with raw poultry. Thoroughly cook poultry, making certain that it is well done. (There should be no pink or red color visible.) Thorough cooking will ensure that all bacteria have been killed. If meat or poultry does not look or smell fresh, do not eat it. If people practice safe handling and cooking methods when using chicken or turkey, it can be a safe and nutritious menu item.

PRUNES ∞ One of the highest fiber sources of all fruits, prunes have a reputation for relieving constipation. In its plum state, the prune's benzoic acid has aided many who suffer from liver disease, and/or blood poisoning.

PUMPKIN SEEDS ∞ The seeds of pumpkins and other winter squashes (such as butternut) are edible. They are abundant in iron and zinc

∞ ∞ ∞

RADISHES ∞ Like many vegetables, radishes come in several varieties, including Asian *daikons*, California whites, and the red globes. The red and whites are sold throughout the year. Black radishes usually become available in the spring and in winter, and they have a longer shelf life. Radishes should be scrubbed and trimmed just prior to using. They are not extraordinarily nutritious but lend a distinct flavor all their own. In Japan, they are sliced with orange sections, and, with a bit of lemon flavoring, they are used as a condiment.

RAISINS ∞ Several varieties of grapes may be dried to become raisins. There are two varieties of raisins, *unsulfured* (black), and *sulfured* (yellow). This fruit has virtually no fat, or cholesterol, and is low in calories. It does contain a fair share of fiber (approximately 1.7 grams for each 3 ounce serving). The sulfured raisins are often the cause of allergy reactions (due to the sulfur content) in individuals who have a sensitivity to allergens. Because raisins are a favorite snack food of children, it is suggested that raisins that have been organically grown are offered. When tested, commercially grown raisins were one of the foods found to have the highest pesticide residues.

RICE ∞ Rice (like beans, grapes, or wines) is more nutritious in the darker varieties. Rice can be combined with nuts, sesame seeds, cheese, or beans to make a complete protein meal. A complete protein includes the eight essential amino acids, those which the body cannot synthesize. Rice can be purchased in many familiar forms, including regular milled, precooked, short, medium, and long grained. High in the B complex vitamins, rice is relaxing to the brain and nervous system. Brown rice contains the most B complex vitamins, as well as iron, protein, and phosphorus.

SEAFOOD ∞ When prepared without the butter and creams, seafood is basically low in calories. Baking or sautéing makes seafood a healthy choice. While many varieties of seafood contain some cholesterol, often this is their source of the omega 3 fatty acids, and these can assist in metabolizing and excreting cholesterol. Seafood is a good source of protein and low in fat. Seafood often contains minimal amounts of vitamin C but several essential minerals, including iron, potassium, calcium, and magnesium. A word of warning, eating contaminated fish can cause an accumulation of mercury within the tissues. Mercury exposure can occur from using fungicides and insecticides containing mercury, which wash down into lakes and rivers, eventually making their way into fish. A tissue mineral (hair) analysis can reveal suspected accumulative amounts and toxicity.

SESAME SEEDS ∞ Sesame seeds are grown from a tall annual plant in India, Africa, and China. Tahini made from them has been given the name "butter of the Middle East." Brought to America with the slave trade, sesame seeds have continued to remain popular in many southern recipes. They are available hulled or unhulled. The unhulled are darker, have the bran intact, and are an excellent source of iron and phosphorus. Although they contain calcium, about half is not available for the body to absorb because it is bound to oxalate. Sesame seeds also contain vitamin T. One ounce of sesame seeds has only 4 grams of fat, and only one gram of that is saturated.

SOYBEANS ∾ *See* **Beans, soy.**

SOY FLOUR ∾ Here is a flour that provides calcium, protein, and various other nutrients. Soy flour can be incorporated into many of our traditional recipes simply by adding it as 1/2 of the total flour called for in the recipe.

SOY MILK ∾ Soy milk can be well tolerated by those with a lactose intolerance or by those who feel that they are not getting sufficient protein in their daily caloric intake. It will complement the diet of those who prefer the vegetarian lifestyle. Those of you who like yogurt, but cannot tolerate the lactose, can look for a product that mimics yogurt but is derived from soy.

SPICES and HERBS ∾ In addition to adding flavor and interest to foods, many kitchen spices and herbs have significant nutritive value and are associated with health benefits, particularly improved digestion. Some of the most popular of these natural flavorings are: **basil** (stimulates appetite, stimulates the stomach, and improves digestion); **caraway seeds** (relieves gas and colic, stimulates appetite); **cardamom** (seeds or powder—stimulates appetite, relieves gas, benefits stomach); **chervil** (improves digestion); **cinnamon** (has been used to settle the stomach); **coriander** (relieves gas); **cumin** (beneficial to the stomach); **fennel** (relieves gas and abdominal cramps); **ginger** (stimulates appetite, relieves gas); **mint** (peppermint or spearmint— benefits stomach and digestion, relieves gas, stimulates bile, acts as a tonic); **parsley** (rich in chlorophyll, vitamins A and C, potassium; relieves gas); **rosemary** (promotes liver function, stimulates bile, aids proper digestion); **sage** (reduces perspiration, tea used as a gargle); **savory** (benefits stomach and digestion, relieves gas); **tarragon** (stimulates appetite, benefits stomach); **thyme** (stimulates appetite, relieves gas; acts as stomach tonic; commonly used in cough preparations); **turmeric** (has anti-inflammatory and anti-bacterial properties, helps to prevent blood clots).

SPINACH ∞ Spinach contains beta-carotene (the precursor to vitamin A), iron, chlorophyll, and an abundant supply of protein, a whopping 12 grams per 100 calories.

SPROUTS ∞ Wheatgrass and alfalfa seeds are two of the most popular grains for sprouting. Harvesting one's own sprouts is a good way to avoid the mold that can easily take place at the supermarket, where sprouts may sit for long periods of time. Sprouting at home is fairly simple. Using a clean glass jar (one quart size), place the seeds inside, after thoroughly rinsing them under warm water. Two tablespoons of seeds to three times as much water seems to be a good mixture. While tinier seeds may require only four or five hours of soaking, others require overnight. The next day, rinse seeds again under warm water, drain thoroughly, and again place inside the jar at about a 70 degree angle, insuring that seeds can properly drain. Store in a dark place. If you like, cover the jar with a clean dry cloth. Seeds must be rinsed and drained twice a day. Alfalfa, cress, mustard seeds, chia, or oat do not require any presoaking. Seeds can be drained by the use of a cheesecloth or a fine mesh screen. Sprouts may be cooked, or served plain. To cook, steam in a pan with a little olive oil and water, not longer than ten minutes.

Alfalfa contains many vitamins and minerals, including chlorophyll. Wheatgrass is similar in nutrients to alfalfa. For more instructions regarding how to sprout various other seeds, check your local library or your computer (Internet resources).

STRAWBERRIES ∞ See **Berries** in this chapter for more general information. In 1992, *Natural Health* reported in their March/April edition that ellagic acid, which is present in cherries and strawberries, counteracts both human and naturally occurring cancer-causing agents.

SUNFLOWER SEEDS ∞ Obtained from the center of the daisy-like sunflowers, these seeds are native to North America. They are high in calcium, thiamine, vitamins B_6 and E, and folacin, but are also abundant in calories and fat. One ounce of sunflower seeds has 14 grams of fat, 2 of which are saturated.

TEMPEH ∞ This soy product is produced by fermenting pre-soaked, cooked soybeans. Another process whereby tempeh is made is through a grain culture known as *rhizopus*. This is achieved by placing a starter culture on hibiscus leaves, then injecting them with hulled soybeans. Tempeh is an excellent source of protein, yielding 31 grams per cup, or 9 grams per each 100 calories. In some countries, such as Indonesia, it is a major staple.

TOFU ∞ Tofu is actually bean curd, made from soybeans. Tofu has a pasty consistency and is high in protein, and because it is rather bland, it will adopt the taste of seasonings and flavorings that accompany it. It can be purchased it a few very different textures, silken, firm, and extra firm. Silken tofu has been used with much success in cheesecake recipes, and firm tofu can be sautéed, or used in casseroles. The extra firm can be marinated and broiled, or baked, along with other recipe ingredients. Some find it a healthy ingredient in the "veggie burger." At an American Cancer Society seminar held in Daytona Beach, Florida, Dr. Stephen Barnes, a biochemist from the University of Alabama, stated that studies involving rats showed that isoflavones, a naturally occurring substance found in soybeans and tofu, seemed to reduce the rate of mammary cancer by half.

TOMATOES ∽ Recent studies have found that tomatoes may be regarded as just as strong an anti-carcinogen as beta-carotene found in carrots, pumpkin, and other vegetables. *Lycopene* (another type of carotene) found in tomatoes is the substance to which this benefit is attributed. Tomatoes are also a good source of vitamins A and C, and are low in calories. They also contain a substance known as *lignin*, which is said to lower cholesterol and prevent gallstones. Tomatoes also provide a source of B vitamins and many of the amino acids. When eating a meal that includes red meat (which creates a good bit of acid in the intestines), tomatoes are a good accompaniment, as they reduce the amount of acid in the intestines at the time of digestion, and they are considered an aid in the cleansing and detoxifying process. The results of a study conducted in Wales showed tomatoes provide protection against certain digestive disorders. Because tomatoes belong to the family of vegetables known as "nightshades," however, they are often the cause of allergic responses in hyper-sensitive individuals.

Tomatoes are quite high in lycopenes, which provide about twice the protection of beta-carotene and may have more anti-cancer abilities. In Italy, a group of researchers conducted a study of lycopene, with tomatoes as the main source. The results of that study were similar to those of other studies—a 40% reduction in the risk of esophageal cancer was shown in the group that ate just one serving per week of raw tomatoes. A 50% reduced rate of cancers of all types was noted among the elderly population of the U.S. reporting a high tomato intake.

TURNIPS ∽ Another of the magnificent 12 cruciferous vegetables, turnips are loaded with vitamins, minerals, fiber, and trace minerals. These vegetables (as stated by researchers) are able to fight free radicals and keep the immune system healthy. Turnips can be steamed (then mashed) with other vegetables, such as yams, celery, apples (and other fruits), onions, and parsnips. If you are allergic to tomatoes, which help the body remove excess uric acid (an acid that, in excess, can cause gout), turnips (or cabbage) can be an excellent alternative.

VEGETABLES ∽ Most vegetables contain vitamins and other nutrients; however, over-cooking or processing for canning may diminish their effectiveness. Thorough rinsing of vegetables may help to remove some of the pesticide residues remaining as well as any bacterial contamination. When available and/or affordable, organic vegetables are recommended. For further information, see entries for specific vegetables.

WALNUTS ∞ Walnuts contain 16 grams of fat, 2 which are saturated.

WHEAT ∞ One of the oldest cultivated grains, wheat contains a fair amount of soluble fiber, B vitamins, and several essential minerals, including magnesium, manganese, and iron. Several years ago, scientists at Cornell Medical Center in New York found that wheat bran can have a positive effect on individuals who have been diagnosed with precancerous polyps of the colon. The benefits derived have been attributed to wheat bran's insoluble fiber content.

YAMS ∞ The yam is a variety of the sweet potato, and it is high in beta-carotene. This vegetable also contains other substances known as *protease inhibitors*, which have been found to have a unique ability to prevent cancer in animals and which may protect the body against certain types of viruses. Some studies using a group of ex-smokers who regularly ate a daily portion of sweet potatoes, squash, or carrots found that they were about 50% less likely to develop cancer after they stopped smoking than those who did not regularly eat these vegetables. According to the National Cancer Institute, the protease inhibitors (along with the carotene) are able to interfere with the processes that allow lung cancer to develop.

YOGURT ∞ When milk is allowed to ferment through a combination of bacteria and yeasts, it gradually transforms into a custard consistency known as yogurt. The milk has its fat removed and is processed with *lactobacillus acidophilus* and other bacteria, which are essential for the health of the intestine.

Because antibiotics destroy good bacteria, often resulting in a condition known as candidiasis, health-conscious individuals would be wise to include this food in their diets. Yogurt is a good source of B complex vitamins and contains a larger percentage of vitamins A and D than does milk. Yogurt also contains protein, calcium, potassium, phosphorus, vitamin B_6, B_{12}, niacin, and folic acid. It contains just as much potassium as bananas. Yogurt is useful in lowering high blood lipid levels and in treating either constipation or diarrhea. Yogurt has been used for many years by herbalists for treating skin diseases and/or kidney problems.

The Long Island Jewish Medical Center reports that eating a cup of yogurt containing the live cultures of lactobacillus acidophilus each day can significantly cut the incidence of vaginal yeast infections. Yogurt has also been found to boost the immune function of animal and human cells, causing these cells to make more antibodies and other disease fighting cells. This occurs through two separate mechanisms that boost the immune system and kill off harmful bacteria. University of Nebraska researchers found that yogurt is more effective as a preventative than a cure for diarrhea and dysentery. Yogurt culture slowed the growth of harmful bacteria by as much as 75% in test mice. The bacteria contained in active yogurt cultures suppress the growth of deadly cancer cells. The enzymes in yogurt suppress the putrefactive (oxidation) of organisms in incompletely digested foods. Yogurt enzymes also prevent gas and bloating. The elderly, with insufficient digestive enzymes, need to supplement their diets with plain yogurt (devoid of yeast and/or sugars), or acidophilus supplements. The National Cancer Institute has determined that malignant tumors shrink in patients who consumed a steady diet of yogurt. Studies at the Harvard Medical School and the University of Chicago Medical School showed that yogurt is especially effective against vaginal cancer. Much of the commercially bought yogurt (with its high sugar content), however, is not thought to be therapeutic because sugar is antagonistic to the B vitamins produced by yogurt bacteria. For vegetarians and lactose-intolerant individuals, there are now many soy yogurts on the market, which can be found in health food stores.

HEALTHIER SUBSTITUTES FOR "EMPTY CALORIE" FOODS

In our resolve to maintain health and avoid disease, it is wise to nurture our awareness regarding the many denatured and devitalized foods. Through my practice, it has become obvious to me that most individuals are usually focused on learning about which nutrients and/or supplements will assist them to correct deficiencies or alleviate bothersome symptoms. However, most seem oblivious to fact that without putting their best effort forward in attempting to avoid certain antagonists or eliminating toxins (many of them known carcinogens), all of the best recommendations in the world will do them little good! More often than not, when it is recommended that clients avoid refined and processed foods, their faces reflect the feelings of a child told not to eat candy. Toxins found in additives accumulate and eventually weaken the immune system, making one vulnerable to disease. Also, most of the naturally occurring nutrients and fiber are removed in the processing of these denatured foods.

This chapter identifies some of the denatured or devitalized foods that may be at the center of one's chemical imbalance or vitamin/mineral deficiencies and offers healthier, tastier food substitutes. Eating healthier does not suggest that taste is the sacrifice one must make! Not at all. It does, however, require that one be willing to begin by instituting certain changes in shopping patterns at the food market. The new food selections are much more nutritious and maybe more satisfying. Those who alter their diets to include healthier foods often begin to notice that it takes less food to satisfy their hunger than it did previously.

The suggested replacements for denatured and devitalized foods are those that will add back the nutrients that were previously lost through refining or processing and will also provide the benefit of reducing saturated fats. This will help to keep unwanted cholesterol and/or pounds down. Healthier food selections will help a body to function more efficiently, and one will begin to experience a more efficient rate of metabolism, with more energy.

∽ ∽ ∽

INSTEAD OF	REPLACE WITH
Supermarket baking powder	Any health food store brand that does not contain aluminum.
Enriched bleached white flours	Organic unbleached white flour, or other whole grains such as spelt, maize, oat, rice, kamut, buckwheat, millet, or soy. Because they are organic (alive!) they must be refrigerated.
Refined white sugar	Rice malt granules, rice malt syrup or barley malt syrup (these sweeteners are 100 times sweeter than refined white sugar and highly nutritious, containing most of the B complex vitamins, which are lost in refined white sugar). For baking, use frozen unsweetened concentrated white grape juice, unsweetened applesauce, date sugar, strained baby food fruits, etc.
Eggs	Eggwhites, tofu, egg substitutes (if one has high cholesterol).
Cream	Canned concentrated skim milk.
Butter or margarine	Butter buds, olive oil, no-fat yogurt, no-fat sour cream, light cream cheese, ghee. Ghee is a clarified butter (fat removed) that can be found in your local health food store.
Coffee	Chicory, herbal teas.
Chocolate	Unsweetened cocoa powder, carob powder.
Cheese	Rice cheese, farmers cheese (low fat), soy or tofu cheese, low fat versions of traditional cheeses.
For sautéing	Use a low-sodium chicken or vegetable broth, or complete the sautéed dish with a cooking sherry.
Salt (sodium chloride)	Garlic, cayenne pepper, onions, herbs, kelp, dulse (sea salts), tamari.
Maple syrup, honey	Bananas, orange juice, extracts.

INSTEAD OF	**REPLACE WITH**
Hamburgers	Vegetable burgers made with soy, seeds, rice, onions, and herbs.
Chili with red meat	Chili with chicken, especially skinless breast meat, or vegetarian chili made with fresh vegetables or a meat substitute.
Baked beans (canned)	Dry black beans, dark red kidney beans; only additional step required is soaking overnight.
White rice, enriched	Wild rice, brown rice, basmati or jasmine rice.
Cold cereals (huge amounts of sugar)	Rolled oats (high in B complex vitamins, phosphorus, amino acids, and fiber.)
White bread	Whole wheat pitas, whole grained breads.
Luncheon meats (high in nitrates, possible carcinogens)	Egg salad, chicken salad, tuna, soup. Some health food stores sell nitrate-free un-processed luncheon meats.

NOTE: The above suggestions are meant to help you get started. As you begin to make the transition to a healthier lifestyle, others will come naturally.

∾ ∾ ∾

FOOD PREPARATION

Following are some tips for promoting healthy eating and food safety:

1. SALADS ∽ Choose the greens, and avoid the eggs, mayonnaise, and other high fat condiments. Store fresh foods immediately after purchasing in order to retain valuable nutrients. Tuna, chicken, or turkey salads often contain mayonnaise, which, due to its high fat content, can be a potential problem when eating out.

2. MEATS and POULTRY ∽ The Centers for Disease Control report that more than six million people get sick each year, and nearly two million lose their lives, due to bacterial contamination (campylobactor and salmonella) and infestation of eggs, beef, pork, chicken, and other types of meat. While the USDA has the job of inspecting these food products for the safety of the consumer, many agree that practices in the last decade have given us much reason for concern. But the following tips can provide some margin of safety:

A. Thoroughly wash and disinfect your hands and cutting board or any other preparation surface prior to and immediately after handling raw meat or poultry.

B. Use a clean dishcloth or towel each time you prepare or cook meat or poultry. You can kill bacteria on dishcloths by placing them in a microwave oven set on high for about 90 seconds.

C. Maintain a refrigerator temperature of 40 degrees or lower and a freezer temperature of zero degrees.

D. Never allow meat or poultry to thaw at room temperatures. Thaw only when ready to use. Thaw in a microwave or in the refrigerator. Should the outside thaw before the inside, the outside will become vulnerable to bacteria, and the meat will take much longer to cook or the food will be insufficiently cooked on the inside.

E. Meat and poultry should be marinated *only in the refrigerator*. Do not risk contamination by placing cooked meat back into the marinade.

F. Never stuff chicken or turkey the night before cooking.

G. Poultry must be refrigerated within two hours of purchase. If it is a warm day, it is wise to pack just-purchased poultry in a cooler until you reach your destination.

3. MILK ∞ Fresh milk usually originates from cows on large dairy farms. The process of pasteurization limits disease, and it is estimated that less than one percent of the millions of quarts of milk consumed in the United States today is unpasteurized. However, even this small proportion of raw milk has become recognized in recent years as a concern, as there are still outbreaks of poisoning, and sometimes fatalities. Though pasteurization requires a heating process to render harmless specific microorganisms that cause infection, this process is not mandatory in all states. In California in 1987 more than 60 people lost their lives to a bacteria that was traced to cheese that was manufactured from unpasteurized milk.

4. EGGS ∞ Raw eggs have been the cause of many salmonella occurrences. It is believed that some of these outbreaks were caused by grade A fresh eggs. It was once thought that this was the result of eating eggs with cracked or contaminated shells. Researchers now believe that the salmonella bacteria from these outbreaks came from inside the hens and bacteria was transferred to the eggs internally. Use only fresh eggs that have been kept stored in a refrigerator. Cook eggs thoroughly and wash utensils and hands if they have come into contact with raw eggs. Make eggnog with a cooked custard, rather than raw eggs.

5. VEGETABLES ∞ Most commercially grown vegetables need to be peeled, washed or scrubbed with a disinfecting solution (such as water with a little added bleach) in order to remove pesticide and fungicide residues. Some products are now available that have been specifically formulated to remove chemical residues and wax from fruits and vegetables. One can make an effective disinfectant from grapeseed extract. Add 15 to 20 drops of Oregon grapeseed extract to 1/2 gallon of water. Shake container for several minutes, and then allow solution to rest for 12 to 20 hours before using.

Chapter IV: Supplements

This chapter provides current information regarding nutritional supplements in concentrated forms, as well as information on several herbs and spices commonly found in the kitchen. Most required nutrients are available to us in dietary food sources, but often our selections are improper. Our basic needs differ slightly or significantly either because of antagonists or certain specific disorders. The USDA has determined that each individual should consume 2 to 4 servings of fruit and vegetables per day, but many are not doing so. Reasons for this might vary, but often people will say that that the produce is not available, they are simply too busy, or that they just do not enjoy eating fruits and vegetables. The benefits of eating fruits and vegetables are many—they provide the daily fiber required to keep the colon healthy and prevent constipation, they lower the total lipids in the blood of individuals prone to high cholesterol levels, they are high in vitamins and minerals, and they provide other benefits as well. People not following a healthy diet may need to use supplements. Many of the concentrated nutritional supplements found in this chapter are considered both natural (extracted from natural sources such as foods) and therapeutic, having been proven in various clinical tests to alleviate many of the disorders caused by deficiencies, genetic requirements, or unusual circumstances. The supplements are listed alphabetically.

ACIDOPHILUS ∾ Acidophilus, or *lactobacillus acidophilus*, is a type of bacterium which ferments milk and has been utilized to treat certain intestinal disorders. by increasing the good intestinal bacteria. Without enough beneficial bacteria, many women are vulnerable to yeast infections. Acidophilus is of benefit to those infected with HIV/AIDS and those who suffer recurrent diarrhea problems. Antibiotic therapy kills the "bad" bacteria, but it also eliminates the "good" bacteria on which we depend for maintaining a healthy intestinal flora. At the Long Island Jewish Medical Center in New York it was discovered that women who ate 6 to 8 ounces of yogurt containing live cultures of *lactobacillus acidophilus* every day, showed a 75% reduction in vaginal yeast infections.

ALFALFA ∾ Alfalfa leaves contain beta-carotene, and its seeds and leaves contain the coagulating factor associated with vitamin K, as well as potassium, phosphorus, iron, and calcium. Alfalfa is said to improve digestion and strength, and it has a good supply of chlorophyll. Alfalfa contains a host of other constituents, such as many of the B complex vitamins, chlorophyll, vitamins C and E, and many minerals. Its root *saponins* have been documented to exhibit selective toxicity toward fungi. Widely used in foods and listed by the Council of Europe as a source of natural food flavoring, alfalfa is listed as GRAS in the United States. Alfalfa is also available in tablets or capsules.

ALMOND ∾ Almond oil is obtained from a small tree in Spain, France, and Italy. It has been reported by researchers to lower cholesterol. In the United States, it is also used as a flavoring agent. This oil is distilled to remove d-hydrocyanic acid, which is very toxic. Almond meal is a highly nutritious powder that can be used as a flour and for making a soothing skin preparation. Because of its low carbohydrate content, it is of value to diabetics. Although somewhat expensive, it can be ordered from a few herb and nutrition suppliers.

ALOE VERA ∞ This product is expressed from the aloe plant leaf. There are over 200 species of this lilylike plant. Aloe contains aloins, anthraquinones, glycoproteins, sterols, saponins, albumin, essential oil, silica, phosphate of lime, a trace of iron, organic acids, and polysaccharides, including glucomannans. It is thought to contain over 50 vitamins, minerals, and enzymes. Cleopatra was said to have used aloe to maintain her beauty. Aloe vera has been used medicinally for over 3,000 years, and it is referred to in the Bible. The most common uses for aloe vera are in healing creams for burns or in solution as an enzyme-producing agent. This ability is most likely attributable to aloe's 0.5% composition of as many as 20 amino acids. It also may be useful as a laxative. It should not be used by individuals using products containing benzoin or balsam, as there may be a cross-reaction. Today it can be found almost everywhere, even in some chain stores and in gallon containers for drinking (in a purified solution).

ANTIBIOTICS (natural) ∞ New information regarding the over-prescribing of antibiotics, the negative effects that persist in the human body after their cessation, the body's changing tolerance levels, and the development of bacterial resistance to antibiotics resulting from overuse have given us good reasons to consider adding natural substances, which may achieve the desired positive results without ill effects. One example of a natural antibiotic that we can add to our diets is acidophilus. A prescribed antibiotic drug may kill unwanted bacteria in our bodies, but it also kills off the friendly bacteria known as intestinal flora. The most important of these friendly bacteria are *Lactobacillus acidophilus* (L. *acidophilus*) and B*ifidobacterium bifidum*, better known as B. *bifidum*. Other forms include L. *brevis*, L. *casei*, L. *cellobiosus*, L. *fermenti*, L. *leichmannii*, L. *plantarum*, and L. *salivaroes*. After any treatment with prescribed antibiotics, it is important to replenish the intestinal flora, and yogurt (minus yeast and sugars) with a live culture of L. acidophilus will accomplish the task. Normal intestinal flora are important to our digestion and the functioning of the immune system, maintaining a healthy environment in our bodies that makes us less vulnerable to bacterial infections. Without this intestinal protection our bodies are vulnerable to such conditions as candida, intestinal tract infections, or even colon cancer. Commonly used therapeutic supplements that act as natural antibiotics include acidophilus, alfalfa, vitamin C (ascorbic acid), garlic, grapeseed extract, hops, horseradish, rosemary, saw palmetto berries, and thyme. *See* Chapter IV, Supplements, and Chapter V, Herbs.

B COMPLEX VITAMINS ∽ *See* Chapter II, Nutrients, under Vitamins.

BEE POLLEN ∽ This supplement is a compound rich in flavonoids, amino acids, vitamins, and minerals. It contains up to 35% protein, and it is said to increase stamina and may be of benefit to those who suffer from allergies.

BETA-CAROTENE (precursor of vitamin A) ❧ Carotenes in general offer a high antioxidant benefit, compared to vitamin A. Antioxidants in carotenes are the factor credited for their anticancer effects, which have been noted in various nutritional studies. Carotenes are also found in carrots, squash, grapefruit, tomato paste, tomato sauce, watermelon, tomato juice, and dried apricots. *See also* Chapter II, Nutrients, Vitamin A.

BETAINE ❧ This compound is used both as a coloring agent and as a dietary supplement. It occurs naturally in beets and in several other vegetables, as well as in some animal substances. It has been used therapeutically with some success in individuals who suffer from muscle weakness.

BETAINE HYDROCHLORIDE ❧ This supplement is employed for those who suffer from digestive and metabolism disorders. Betaine hydrochloride is a source of hydrochloric acid, the main acid responsible for digestion and assimilation.

BILBERRY ❧ Bilberry is an enzyme extracted from the darkest of blueberries grown in the Alps and in Scandinavia. Several recent studies have demonstrated that bilberry (*anthocyanins*, or the blue-colored chemicals) given orally improves vision in healthy people and also helps to treat others with certain eye disorders. Anthocyanins act to prevent vessel fragility and blood clot formation. It has further been reported that bilberry increases prostaglandin release from arterial tissue, which dilates blood vessels. Bilberry also contains a compound known as *arbutin*, a diuretic and anti-infective agent derived from the dried leaves. Recently, arbutin has been recommended (along with other nutrients and minerals such as vitamin C and zinc) for the treatment and/or prevention of cataracts and other eye diseases.

BIOFLAVONOIDS ∽ Also referred to as vitamin P, bioflavonoids are a group of brightly colored, water-soluble substances found in many fruits and vegetables, along with vitamin C. Michael Murray, N.D., a faculty member of Bastyr University, where alternative medicine is studied, states that taking vitamin C with a product that contains a meaningful amount of bioflavonoids can benefit nutrient absorption and cause a general sense of well-being. With other compounds such as rutin, flavones, and flavonols, bioflavonoids also have an added benefit in that they work like an antihistamine. Also, they reportedly enhance the integrity of blood vessels.

BIOTIN ∽ Part of the B complex. More information on biotin is available in Chapter II, Nutrients.

BLACKSTRAP MOLASSES ∽ Blackstrap molasses is a food with one of the most abundant supplies of vitamins and minerals. It comes from sugar cane that has not yet been refined. Blackstrap molasses contains more calcium than milk, more iron than many eggs, and more potassium than any food. It is a rich source of the B vitamins and is easily absorbable by the digestive tract. Also, it contains large amounts of copper, magnesium, phosphorus, pantothenic acid, inositol, and vitamin E. Molasses may be used as a sugar substitute in cereals and may be eaten instead of jam or jellies. Just one tablespoon of blackstrap molasses contains 3 milligrams of iron and over 100 milligrams of calcium. Varicose veins, arthritis, ulcers, hair damage, eczema, psoriasis and dermatitis, as well as angina pectoris, constipation, colitis, anemia, and nervous conditions have responded well to supplementation of blackstrap molasses. *See also* Chapter III, Dietary Sources, Molasses.

BREWER'S YEAST ✎ This single compound probably contains more essential nutrients than does any other supplement. It contains all of the B complex vitamins, and approximately 40% is protein. It also contains many organic minerals, such as zinc and iron. Brewer's yeast also provides most, if not all, of the essential trace minerals. In addition, brewer's yeast is one of the best sources of RNA, a nucleic acid that is extremely beneficial in helping maintain a strong immune system and in the prevention of degenerative diseases. This food also contains 16 amino acids and is abundant in the mineral phosphorus. It is recommended that for each tablespoon of brewer's yeast, 8 ounces of skim milk or 4 tablespoons of dry powdered milk be taken at the same time, in order to keep the calcium-phosphorus balance. For dosage, follow instructions as given on the supplement bottle, according to potency and age of individual. Brewer's yeast is easily digested and assimilated.

BROMELAIN ✎ This is an enzyme that is protein digesting. Bromelain is found most abundantly in pineapple. Taken orally, it has been shown to be beneficial in the treatment of inflammation and swelling of soft tissues often associated with traumatic injury. Taken in combination with *quercetin* (from the bioflavonoid family), it is found to effectively block some common allergic responses. (*See* **Quercetin.**) These two compounds have the unique ability of lightly coating the cells known to release histamines, which the body releases in response to allergens and which cause such annoying symptoms as sneezing, runny nose, and itching eyes.

CALENDULA ∽ Also known as marigold, this flower's head contains essential oils, carotenoids, saponin, resin, and a bitter principle. Herbalists and traditional pharmacists use this oil to treat minor skin irritations. Marigold flowers have been reputed to lower blood pressure but may be harmful to pregnant women. Calendula has been employed by herbalists to treat fevers, ease menstrual cramps, and ease eruptive skin diseases; it is available as an extract.

CAPSICUM ∽ Native to tropical South America, capsicum, another name for peppers, has been used medicinally by herbalists and modern pharmacologists as an internal disinfectant and for easing the pain associated with arthritis. It is also said to enhance circulation and manage any excess *fibrin* (the clotting factor in the blood), which could contribute to a stroke.

CAROTENE ∽ A yellow pigment occurring in yellow vegetables, egg yolk, and other food, carotene is converted by the body into vitamin A. *See* **Beta-carotene** and Chapter II, Nutrients, Vitamin A, for more information.

∽ ∽ ∽

CASTOR OIL (castor bean or alphamul) ∞ Originally native to India, the castor oil plant is one of the most ancient sources of medicine known. Castor oil seeds have been found in Egyptian tombs thousands of years old. The oil, extracted from the bean, contains *glycerides* of *ricinoleic, isoricinoleic,* and *stearic acids*. The plant was mentioned by Hippocrates, circa 400 B.C. Herbalists have used castor oil to treat food poisoning, to expel worms, and topically to dissolve cysts, growths, and warts. Medical doctors have long recommended castor oil as a stimulant and/or as a laxative.

CAYENNE ∞ Extracted from the pungent fruit of the pepper plant, cayenne originated in Central and South America, where it was used by natives for many diseases, including diarrhea and cramps. It contains *capsaicin, carotenoids, flavonoids,* essential oil, and vitamin C. Scientists report that cayenne prevents the absorption of cholesterol. Many herbalists report that it can lower blood pressure. Cayenne is non-irritating when uncooked, and its powder or tincture can be rubbed on toothaches, swellings, and inflammations. Cayenne is available in capsule form.

CHLOROPHYLL ∞ Known as the green coloring matter of plants, chlorophyll plays an essential role in the photosynthesis process. Used topically, it aids in promoting normal healing and relieves pain, inflammation, and skin irritations. It is employed orally for reducing odors, such as those associated with colostomy, ileostomy, and incontinence. High dosages may cause a photosensitivity in some individuals.

CHOLINE ∽ *See* Chapter II, Nutrients, Vitamins.

CHONDROITIN SULFATE ∽ Chondroitin is a major constituent of cartilage in the body, and chondroitin sulfate is a
supplement form which seems to slow down enzymes that
contribute to the breakdown of cartilage and connective tissue. Studies suggest chondroitin sulfate may promote formation of new, healthy cartilage by stimulating cartilage
cells. Clinical research shows that when chondroitin sulfate is taken on a daily basis, positive changes begin to happen in injured joints; 400 milligrams of pure, high-quality
bovine chondroitin sulfate is available in capsule form. It is
best to consult with a professional when considering any supplementation. Duration, dosage recommendation, and other
considerations are all part of each individualized protocol.

CHROMIUM ∽ This mineral is involved in the breakdown of
sugar for conversion to energy. *See* more on chromium in
Chapter II, Nutrients, Minerals.

CLOVER, RED ∽ As a constituent of herbal teas, red clover
is a perennial plant common throughout the United States
and Europe. Its constituents are carbohydrates, *coumarins*,
isoflavonoids, *saponins*, *tannins*, and other compounds. It is being reviewed pharmacologically and in animal studies as a
possible chemo-protective agent. This effect has been documented for its *biochanin* (an *isoflavonoid*) content. No known
toxicity has been discovered in the use of red clover; however, excessive ingestion is not recommended, so that possible imbalances are avoided. It is available as an extract
and in pill or capsule form.

COENZYME Q10 ∽ Also known as *ubiquinone*, coenzyme Q10 is an essential component of the *mitochondria*, the energy producing unit of the cells in our body. Professor F. L. Crane and his colleagues at the University of Wisconsin first discovered coQ10 in 1957. Continued research has been conducted by Dr. Karl Folkers of the University of Texas. Research now indicates that coQ10 supplementation can provide significant benefit as an antioxidant and for the treatment of a number of health disorders. Recently, it was reported that a woman given only three weeks to live by her medical physician, due to her degenerative heart condition, began supplementing with coenzyme Q10 on the advice of an alternative health provider. That was three years ago, and today the woman is reported to feel great! Coenzyme Q10 is said to enhance oxygen availability, which increases cell energy. Other benefits associated with the supplementation of coQ10 are to lower LDL (bad cholesterol), increase energy, strengthen the immune system, and protect the cardiovascular system.

COLLAGEN ∽ An insoluble protein found in connective tissue, including skin, bone, ligaments, and cartilage, collagen represents about 30% of total body protein. Collagen is hydrolyzed through dilution with water, turning it into another substance (such as when starch is mixed with water, then turned into glucose): hydrolyzed collagen supplements are just that. A collagen supplement is intended to support hair, nails, muscles, tendons, cartilage, ligaments, bones, gums, teeth, eyes, and blood vessels.

∽ ∽ ∽

COUMARIN (tonka bean) ∞ This fragrant ingredient of ton-ka beans is made synthetically as well. It was used in com-bination with other compounds pharmacologically as an anti-clotting agent, especially for individuals who may have an artificial valve replacement. Today the FDA prohibits coumarin in foods, because it is carcinogenic and toxic by ingestion. It is not to be confused with cumin.

CRANBERRY JUICE ∞ In 1984 *The Journal of Urology* revealed that cranberry juice is an excellent inhibitor of bacterial ad-herence in the urinary tract. It has long been considered an in-testinal antiseptic, possibly attributable, in part, to its high vi-tamin C content. The sugarless concentrate is recommended.

CURCUMIN ∞ *See* **Turmeric.**

CURRANTS ∞ The small red and white berries are cultivated in Greece and found by many herbalists to be antiseptic and effective against many blood disorders as well as jaun-dice, skin conditions, and liver disorders. Black currants are used to treat infant thrush. Individuals with bladder problems and pain caused by urinary retention may find relief in drink-ing an infusion made from freshly gathered black currants. Currants are used in cough and sore throat preparations.

DANDELION ∞ *See* Chapter V, Herbs.

DHEA (dehydroepiandrosterone) ∞ DHEA is a hormone man-ufactured by the adrenal glands. In 1994, researchers at the University of California, San Diego School of Medicine, wanted to find out what would happen to older individuals if they took supplements of DHEA. Middle-aged volunteers were given 50 milligrams of DHEA nightly for a period of three months. Participants related that while on DHEA, they felt a physical and psychological enhancement. Subjects reported more energy, better sleeping patterns, improved mood, and increased ability to deal with stressful situations. When the results of this study were published, there was a rush of excitement, including more publications pointing to DHEA as "the supplement of the month."

The list of potential benefits associated with the supplementation of DHEA is lengthy. It is said to improve immunity, to turn on "youth" genes in our DNA that might be shut off by low levels of DHEA, to decrease the rate of cancer in individuals, to lower cholesterol levels in the blood, to improve insulin's function, and to provide a host of other benefits. However, no long-term studies have been done, thus the long-term safety of this hormone is yet to be determined. According to Dr. Michael Bennett, from the Department of Pathology at the University of Texas, Southwestern Medical Center in Dallas, who has been researching DHEA's effects on the immune system, DHEA does not have a bad effect such as those caused by androgenic steroids, often used by body builders. DHEA is a precursor of estrogen and can possibly lead to breast enlargement in men if used in high dosages. In women, high dosages could pose the risk of an over-abundance of androgens, leading to such side effects as *hirsutism* (excess body hair). Male mice lived longer when given supplements of this steroid. Dr. Alan Gaby, a leading nutritional expert, believes that DHEA has a role in hormone replacement therapy. His studies have found it beneficial for anti-aging, increasing muscle strength, and building bone density.

Because over-the-counter supplementation of DHEA is relatively new, no specific dosage standards have been set. Just as with many supplements, dosage requirements vary with each individual. Individuals who may be at greater risk of negative symptoms associated with the use of DHEA include individuals under age 40, pregnant women, nursing mothers, and those who could be pregnant. Also, it is recommended that, prior to taking DHEA supplementation, you should have your present DHEA levels determined through a saline test. Studies of drug interactions have not yet been done; however, according to some physicians, you should not take DHEA if you are taking aspirin or blood thinners, stimulants (including herbs), or thyroid medications. Side effects with high dosages of DHEA include acne, hair growth in unwanted places in women, irritability or mood swings, overstimulation or insomnia, and fatigue or low energy levels. My conclusion is that the use of DHEA should be monitored by a professional, and it is not suggested that this hormone be used in any self-treatment plan. (*See also* **Hormones** in this chapter.)

EVENING PRIMROSE ∽ The leaves and oil from the seed of the evening primrose are used by herbalists to treat liver and kidney dysfunctions. The oil is high in essential fatty acids (linoleic and linolenic) of the type which are converted into prostaglandins and hormones. It is utilized for lowering blood pressure in some individuals and may also be useful in some types of inflammatory conditions.

∽ ∽ ∽

FIBER ∞ Fiber is non-digestible (insoluble) food content that adds roughage to one's diet. It is beneficial in stimulating the bowels, thereby preventing constipation. Constipation of long standing has been linked in some cases to colon cancer. Insoluble fiber is derived primarily from grains. It is available in two specific forms, cellulose and hemi-cellulose, as well as in some forms of lignins. Soluble fiber, derived mostly from fruits and vegetables, contains lignin, pectin, and gums. A healthy diet should contain a combination of the two.

FISH OIL ∞ *See* **Omega 3 fatty acids**. In the spring of 1993, Dr. Rashida Karmali, Associate Professor of Nutrition, Rutgers University, announced findings that supplements of fish oil equal to what Japanese women commonly eat in fish suppressed biological signs of developing breast cancer in cancer-prone women.

FLAVONES ∞ Derived from certain groups which range in color from ivory/white to yellow, flavones are thought to stimulate the heart and reduce fluid accumulation. Some of these flavones (*hesperidin* and *rutin*) are thought to also lower blood pressure and strengthen the immune response. *See* **bioflavonoids**.

FLAX (linseed) ∞ Containing essential fatty acids (EFAs), flax goes back as far as pre-historic times, where it was first cultivated in Mesopotamia. This nutrient has a list of benefits that are almost more numerous than any other supplement recommended by today's health consultants. *See* Chapter II, Nutrients, Essential Fatty Acids.

FOLIC ACID ∼ Also known as vitamin B_9, folic acid is part of the vitamin B complex and is needed for cell growth, red blood cell formation, and protein metabolism. It is used for the treatment of folic acid deficiency caused by alcoholism, as well as for anemia, diarrhea, fever, hemodialysis, liver disease, or surgical removal of the stomach. A deficiency in pregnancy has been linked to several birth defects, including neural tube defects such as spina bifida. Deficiencies of folic acid are also associated with such conditions as depression, osteoporosis, and atherosclerosis. It can be found abundantly in the dark leafy vegetables, such as spinach, mustard greens, and cabbage. A good supplement for folic acid and all of the other B complex vitamins is brewer's yeast. Folic acid can also be found in fair concentrations in wheat germ, liver, kidney beans, lima beans, black-eyed peas, asparagus, lentils, and walnuts. A 3 1/2 ounce serving of the above will provide 105 micrograms of folic acid from lentils and 440 for black-eyed peas. Brewer's yeast can provide a whopping 2,022 micrograms of this substance. For more information on folic acid, *see* Chapter II, Nutrients Vitamins.

FRUCTOSE ∼ Researchers at the General Clinical Research Center at the University of Colorado School of Medicine, Denver, have found that fructose (a sugar derived from honey and various other fruits) is absorbed into the gastrointestinal tract more slowly than other sugars. In combination with dextrose and phosphoric acid, it is useful for treating vomiting and nausea.

∼ ∼ ∼

GAMMA-ORZANOL (esters of ferulic acid) ∽ Found in grains and isolated from rice bran oil, ferulic acids are compounds found in nature, and ferulic esters are available in wheat, barley, oats, asparagus, rice, berries, tomatoes, citrus fruits, peas, and many other foods. The bran portion of whole grains contains the largest amount of this substance. Ferulic acid is ten times greater in whole wheat than in white flour (500 micrograms per gram in whole wheat, and only 50 micrograms in processed white flour). It has been used medicinally by the Japanese for more than 25 years. It has been shown to be beneficial for the treatment of minor anxiety disorders, menopausal effects, and high cholesterol levels. Because of gamma-orzanol's high antioxidative effects, it is now being reviewed as a therapy for preventing some of the negative side effects often associated with radiation or chemotherapy. Also, in laboratory studies with animals, gamma-orzanol appears to provide potent anticancer effects. Gamma-orzanol is available at health food stores in capsule form. When taken orally, it is converted into ferulic acid. The ferulic acid availabilities as contained in gamma-orzanol are evidently the most useful form of this antioxidant.

GARLIC (Allium sativum) ∽ Garlic may be most hailed for its benefits during World War I, when it was used in the treatment of typhus and dysentery. Medicinal use goes all the way back to the time of Hippocrates, when he used garlic to treat infected wounds and pneumonia. The renowned Albert Schweitzer used garlic effectively against typhus, cholera, and typhoid. Garlic contains high amounts of fluorine, sulfur, potassium, phosphorus, and the vitamins A and C. More recently, it has been found to contain antibiotic, antiviral, and antifungal properties. Many recent studies confirm that garlic may lower blood pressure levels and decrease the ability of platelets to clot, possibly preventing strokes. In some scientific literature, it is said that garlic may decrease nitrosamine formation (a powerful cancer-causing agent created when nitrates combine with natural "amines" in the stomach). In 1992, at the American Chemical Society meeting in Washington, D.C., three Rutgers University researchers reported that chemicals in garlic may protect the liver from damage caused by large doses of the painkiller, acetaminophen, and may prevent the growth of lung tumors associated with tobacco smoke. Other studies report that garlic can lower blood serum cholesterol and triglyceride levels. For more information on garlic, *see* Chapter III, Dietary Sources. Garlic is available in tablets or capsules.

GLUCOSE ∞ Glucose is sweeter than sucrose and occurs naturally in blood, grapes, and corn. Glucose is used therapeutically in the treatment of diabetic coma as well as for soothing the skin.

GRAPESEED ∞ Grapeseed is a powerful antioxidant and antibiotic. It can protect the body from free radicals. As a doctor of naturopathy, I have often recommended it to the HIV/AIDS client for assistance in maintaining a proper internal pH balance. Individuals prone to candida or thrush may also be able to benefit from this supplement. Some studies have shown that it can be utilized (in an extract form) to purify drinking water. When I've consulted with clients anticipating traveling and seeking a way to avoid travelers' diarrhea, I have often recommended grapeseed extract: 10 drops in a half gallon of water left to sit for 12 hours or so and then shaken thoroughly is a beneficial recipe for using this plant. For more information, *see* Chapter III, Dietary Sources.

GREEN TEA ∾ Tea leaves come from an evergreen of the camellia family, "Theaceae," principally cultivated in China, Japan, Sri Lanka, and other Asian countries. The leaves are, however, processed in very different ways, and it is the processing itself that is responsible for the various colors and tastes of the specific teas. Each emerges with an entirely unique chemical structure. Black tea is more popular to date with most Americans, and is heated to temperatures of 200 degrees, or higher, for many hours. Oolong tea is heated less than black tea. Green tea is not heated, but is slowly steamed for a predetermined time, which makes it quite different from the black teas. Among tea drinkers, green tea has been found by scientists to provide the most positive effect on health, attributed mainly to the *polyphenols* contained in green teas, more than any other type of tea. Polyphenols behave much like antioxidants in combating destructive compounds, or free radicals.

GUAR GUM ∾ Pronounced gwar, this substance is derived from a nutritive seed tissue of guar plants cultivated in India. It is made from cluster beans and used chiefly as a thickening agent in processed foods. Recent scientific studies conducted over the past ten years report that guar gum may be significant in lowering serum cholesterol levels.

A study done by Finnish researchers showed that guar gum significantly lowered LDL (or bad) cholesterol levels. Guar gum is available in capsules or in combination with other sources of soluble fiber.

HOPS (*Humulus lupulus*) ∞ The name "hops" generally pertains to the scaly, cone-like fruit that develops from the female flowers found on this perennial climbing plant. Extracts of hops have been utilized for the treatment of various diseases, including pulmonary tuberculosis, leprosy, and acute bacterial dysentery. Two bitter acids in hops, *lupulon* and *humulon*, have been shown to kill bacterial strains of staphylococcus. Several studies indicate an estrogenic activity in hops comparable to, or even higher than, the daily intake of estrogens found in some oral contraceptives. Hops is used in supplements to to induce relaxation and sleep. *See also* Chapter V, Herbs.

HORMONES ∞ Previously, there has not been a need for evaluating the potential use as a supplement of individual hormones, since consumers have not been exposed to hormones other than the traditionally prescribed ones utilized for the past 30 years, such as contraceptive drugs and hormone replacement therapy (HRT) for menopausal women. Today, however, there are new hormone supplements, such as melatonin and DHEA, available to virtually anyone without a prescription. Because these supplements are sold over the counter at many pharmacies and health food stores across the nation, individuals often believe they are harmless. In addition, the enticing advertised claims for these products are not based on biochemical studies or clinical trials.

It would be wise, then, to take this information into consideration when contemplating the use of these drugs.

To better understand the correct use of hormone supplements, we must look at the role of hormones in the body. The purpose of the endocrine and nervous systems is to create and maintain homeostasis, or internal harmony. While these two systems can work singly or together, as in the *neuroendocrine* system, they differ in the ways information is relayed through the body. The nervous system uses a series of electrical impulses to transfer information quickly to two kinds of cells, muscle and gland. The endocrine system uses hormones to chemically transfer information, creating slower and longer lasting effects. Most hormones affect specific parts of the body, even though their form of transport is through the bloodstream. Hormones target specific cells and cause them to become stimulated and to react in a particular way. For example, growth hormone, or *somatotropin*, is thought to affect bodily growth indirectly by accelerating the transport of amino acids into cells. Evidence seems to substantiate this. It has been noted that the blood content of amino acids decreases within hours after administration of growth hormone to a fasting animal. The entrance of amino acids into cells is made faster, and the rate at which the amino acids form tissue proteins accelerates when somatotropin is added. This tends to promote cellular growth. Growth hormone stimulates growth of both bone and soft tissues.

Other hormones have other specific responsibilities; some hormones of the brain's anterior pituitary gland secrete the hormone generally considered the most important for maintaining human pigmentation, *melanocyte-stimulating hormone*, or MSH. Neurons in certain parts of the brain synthesize chemicals, which are secreted into the blood. Other hormones, the four tropic hormones secreted by the anterior pituitary gland, all have a profound effect upon the thyroid, pituitary, and endocrine glands. While all hormones work to accomplish basically the same task, each is responsible for stimulating another into action, much like a chain-reaction effect. *Thyrotropin* promotes and maintains growth and development of its target gland, the thyroid, and stimulates the production of thyroid hormone. *Adrenocorticotropin* (ACTH) promotes and maintains normal growth and development of the adrenal cortex and stimulates it to secrete *cortisol* and other *glucocorticoids*. *Follicle-stimulating hormone*, or FSH, also stimulates primary ovarian follicles to start growing and to continue developing to maturity, that is, to the point of ovulation. FSH stimulates follicle cells to secrete estrogens, one type of sex hormone. In males, FSH stimulates development of semen. *Luteinizing hormone*, or LH in the female and ICSH in the male, as its name suggests, functions by stimulation of the *corpus luteum*: their ultimate responsibility is to stimulate estrogens in the female and progesterone in the male. The posterior lobe of the pituitary gland is also a storage and release area for two hormones known as *antidiuretic hormone* (ADH) and *oxytocin*. The thyroid gland secretes the thyroid hormone *calcitonin* throughout our bodies. Each hormone is dependent upon the stimulation of another hormone for its reaction or response. A deficiency of certain hormones can cause disorders. Too little growth hormone, for example, may cause dwarfism; too much may cause giantism. An excess of certain hormones can also lead to disorders. For example, Graves' disease and goiter are caused by hypersecretion of the thyroid gland.

Supplementing hormones that are not deficient can upset the body's delicate internal balance of hormones, which may lead to other conditions, symptoms, and disorders caused by the hormonal imbalance. Some of these conditions can be very serious. Biochemistry is a science best left to the experts. Many of the synthetically manufactured hormone supplements have not yet been studied enough to be regarded as both safe and effective. Two of the most publicized and frequently sold hormone supplements on the market today are *melatonin* and DHEA, or *dehydroepiandrosterone* (pronounced D-hi-dro-epp-E-andro-stehr-own). They are discussed within the alphabetical listings in this chapter.

HORSERADISH ∞ *See* Chapter V, Herbs.

KELP ∞ Obtained from the Pacific marine plant M*acrocystis pyifera*, kelp contains many minerals, including a good supply of iodine, which is probably why so many herbalists use it to treat mild thyroid deficiencies. Perhaps it also works well because the iodine content is organic and therefore most compatible with the organic body, easily assimilated, and best utilized. The Japanese have long used kelp for burning excess body fat, and it is also believed to help regulate the texture of the skin. Kelp can be found at your local health food store in a small shaker bottle and is often substituted for sodium chloride (table salt). It is relatively inexpensive.

If headache develops, reduce intake and then resume use. *See also* Chapter III, Dietary Sources.

LACTOBACILLUS ACIDOPHILUS ∽ *See* **Acidophilus.**

LECITHIN ∽ This is a phospholipid, which, along with proteins and carbohydrates, make up the main structural components of cells. The lecithin made from pure soy is a good source of EFAs (essential fatty acids). Lecithin also contains many B vitamins, such as choline and inositol, as well as phosphorus and linoleic acid. People with hereditary atherosclerosis may have abnormally high genetic requirements for particular nutrients such as linoleic acid that properly metabolize fats. Lecithin has been shown to have the ability to break down fats in the blood. Lecithin and choline, also found in egg yolk, are what help keep the cholesterol moving in the bloodstream.

LIPOIC ACID (also referred to as thiotic acid) ∽ A sulfur-containing, vitamin-like substance, lipoic acid plays an important part (as a cofactor) in two very vital reactions in the production of cellular energy. While the healthy body is capable of manufacturing this substance, under specific unusual circumstances, a deficiency may occur. Lipoic acid is somewhat unique, in that it is effective as an antioxidant both in a water- or fat-soluble environment. This makes it very useful as an anti-cancer combatant. In Germany, lipoic acid is an approved drug in the treatment of diabetic neuropathy, a degeneration of the nerves. Some studies indicate that HIV/AIDS infected patients could also benefit from supplementation of this compound. These studies show that lipoic acid can greatly inhibit replication of HIV.

Lipoic acid works well with other antioxidants, such as vitamins C and E. It may improve blood sugar control in diabetics, and thus allow a reduction in the dosage of insulin or sugar-lowering drugs. Dosages as a general antioxidant range from 20 milligrams to 50 milligrams daily. In the treatment of diabetes, the recommended dosage may be from 300 to 600 milligrams daily, while in the treatment of AIDS, the usual recommended dosage may be 150 milligrams. There are no readily available reports of any negative symptoms associated with the supplementation of lipoic acid, as reported in over 30 years.

MELATONIN ∞ When melatonin came on the scene and received rave reviews in the August issue of Newsweek in 1995, it suddenly became an overnight success as a super supplement. Melatonin is said to alleviate insomnia, treat jet lag, reduce stress (and depression), fight cancer, enhance the immune system, prevent heart disease, and behave much like an antioxidant. After Newsweek's article, there were thousands of people flooding the health food stores hoping to secure this miraculous supplement. Melatonin is a hormone that our bodies make from serotonin, which is secreted by the pineal gland. The exact function of melatonin is still widely misunderstood. It is thought to be a major contributor in the synchronization of hormone secretion, in that it is part of that internal regulator often referred to as our biological clock. Melatonin is less available in the body in the later part of the day, because the brain is secreting less of this hormone; therefore, melatonin is useful for temporary treatment of insomnia and jet lag. It is thought to also be beneficial in some types of depression, and is often recommended to cancer patients to assist them in securing a restful night's sleep.

Since this is a hormone that decreases with age, most individuals under age 40 need not be supplemented with melatonin. In any case, the goal should be to determine the cause of insomnia and attempt to correct that, rather than induce sleep. At this time, there are no recorded cases of toxicity or serious side effects associated with proper use of melatonin; however, long-term studies are incomplete. Dosages recommended through supplementation are usually more than the 24-hour urinary excretion rate for melatonin, which is about 0.03 milligrams. Dosages of 0.1 milligrams and 0.3 milligrams can stimulate a sedative response when melatonin levels are low. As an adjunct therapy in cancer patients, higher doses of as much as 3 milligrams at bedtime may be required. According to Michael T. Murray, N.D., co-author of *Encyclopedia of Nutritional Supplements*, in one study, a daily dosage of 8 milligrams for only 4 days resulted in significant alteration in *circadian rhythm* (the regular sleep rhythms) through a dislocation of day and night caused by high speed travel. Because vitamin B_{12} can influence melatonin secretion, low levels of melatonin in the elderly could be the result of low vitamin B_{12}.

More research into the effects of melatonin and other hormones must be conducted, including long-term studies, before we can fully understand their mechanism of action. Therefore, caution is advised when one is considering melatonin supplementation. Consult with your physician or naturopath first. (*See also* **Hormones** in this chapter.)

MOLASSES ∞ See **Blackstrap molasses**.

NIACIN ∞ Part of the B complex. (See Chapter II, Nutrients, Vitamins.)

OAT FIBER ∞ Rich in mucilage and B vitamins, oat fiber is also available in an extract form for treating anxiety. Recent research has shown that in its flour form (from the grain), it is beneficial for lowering cholesterol. In ointment form, it is useful for relieving the symptoms associated with skin irritations, such as itching. One can also purchase oatmeal soap (Avena sativa).

OLIVE OIL ∞ Mentioned often in the Bible, the olive tree has provided us with many benefits throughout the centuries. The oil has been used as a laxative, and the leaves have been helpful for reducing fevers or as a mild tranquilizer. More recently, olive oil has been found to be effective in lowering LDL (bad cholesterol) levels without affecting the HDL (good cholesterol levels). See also oils in Chapter III, Dietary Sources.

OMEGA 3 FATTY ACIDS ∞ These oils, found in fish, are reported to lower fats in the blood, and therefore, lower the incidence of coronary artery diseases. See EFAs (Essential Fatty Acids) in Chapter III, Dietary Sources.

PAPAIN ∞ An enzyme prepared from the papaya, papain is a potent digestant of dead protein matter, but is harmless to live tissue. This same extract is used in meat as a tenderizer. See **Papaya**, below.

PAPAYA, PAPAW, or MELON-TREE ∽ Herbalists use the fruit, seeds, and leaves of this purple flowered tree to ease menstrual problems and cramps. In 1992, researchers at Purdue University isolated a powerful anticancer drug from the papaw tree. In animals, it has proven to be many times more potent than the widely used cancer drug *adriamycin*. Compounds obtained from the bark of the papaw tree in North America seem to shut down the power supply of cancer cells, rendering them unable to grow. Scientists seem to feel that this compound might be utilized with many positive results for cancer cells now resistant to chemotherapy.

PECTIN ∽ Pectin is a soluble fiber found in roots and fruits, most commonly in lemon and orange rind. Pectins are presently used in anti-diarrheal medications. In the late 1950s, researchers reported that dietary pectin increased the excretion of fats, cholesterol, and bile acids. Another group of researchers found that adding pectin to a meal lowered blood sugar in non-diabetics as well as non-insulin-requiring patients. Pectin is also the basis for fruit jellies.

PEPPERMINT ∽ *See* Chapter V, Herbs.

POTASSIUM ∽ *See* Chapter II, Nutrients, Minerals.

PSYLLIUM ∽ Used as a laxative since the 1930s, psyllium absorbs water to increase the bulk content of the stool and stimulates movement. Today, it is also used for appetite control (in a capsulated form), when taken with water about an hour before each meal, and as an aid to help lower blood cholesterol levels. Taking this and similar supplements such as guar gum in a capsule form with plenty of water (suggested 16 oz.) will prevent any expansion of the compound, which could cause the substance to swell in the esophagus, prior to reaching the stomach, and lead to adverse reactions, such as breathing problems.

PUMPKIN SEED ∾ Due to the pumpkin seed's large organic zinc content, along with a few of its other ingredients such as lecithin and phytosterol, it is highly touted today as an effective treatment for *prostatitis* (inflammation of the prostate gland).

QUERCETIN ∾ Found mainly in plants, quercetin is a bioflavonoid usually extracted from the inner bark of an oak tree native to North America. Its active compound, *isoquercitin*, has been found to block allergic and inflammatory reactions by inhibiting an antibody responsible for the common varieties of allergic responses and by inhibiting *leukotrienes* and *prostaglandins* involved in producing inflammation. Bioflavonoids are a group of water-soluble, brightly colored compounds often available in fruits and vegetables. Without bioflavonoids, the benefits derived from vitamin C might be lost or unavailable. Also, bioflavonoids provide an anti-histamine effect. Like many other concentrated nutrients, bioflavonoids have many other benefits. They strengthen the integrity of the arterial walls and may lower the excess fibrin in the blood that has frequently been associated with heart attacks and strokes. Some types of bioflavonoids are believed to also work as antioxidants.

ROSE HIPS ❧ Rose hips are the fruit of various species of wild roses. The extract form is employed as a natural food flavoring and as a tonic. Rose hips are reputed to correct constipation and to fight off the common cold. Rose hips are rich in vitamin C. Some vitamin C supplements are made from rose hips.

ROSEMARY ❧ *See* Chapter V, Herbs.

ROYAL JELLY ❧ Honey is the common sweet material taken from the nectar of flowers and manufactured in the sacs of various kinds of bees. Honeybees also produce royal jelly, which they feed to queen bees, hence its name. It contains many vitamins and nutrients, including an unknown compound that may be useful to individuals who suffer from allergies.

RUTIN ❧ Found in many plants, especially in buckwheat, rutin is used as a dietary supplement for strengthening the blood vessels.

SAFFLOWER ❧ Cultivated in various parts of Europe and America, safflower is native to Egypt. Widely used in herbal teas today, it is said to be a fever-reducing compound. In concentrated doses, it will have a laxative effect.

SAW PALMETTO ❧ Frequent as well as chronic prostate infections have been shown to respond favorably to a concentrated, or extract, form of saw palmetto berries. Saw palmetto contains invert sugars, fixed oils (including *oleic*, which is unsaturated, *capric acid*, *caprylic acid*, *palmitic acid* and a saturated *stearic acid*), steroids, and flavonoids. Two double-blind studies documented that those individuals receiving saw palmetto, as opposed to the placebo, responded with significant improvement in their ability to increase urine flow and subsequent shrinkage of the prostate gland. Saw palmetto has also been proven useful for some cases of problems with sterility, bronchitis, kidneys, and reproductive organs. Studies have reported saw palmetto to be well tolerated, and the results of standard blood chemistry tests in patients using it were normal. *See* Chapter V, Herbs.

SELENIUM ∽ This essential mineral is also a potent antioxidant. See more on selenium in Chapter II, Nutrients.

SHITAKE MUSHROOM ∽ This dark Oriental mushroom is employed by many herbalists today for many reasons. The shitake mushroom may be an important food or supplement for everyone. In China it is classified as a superior herb. Naturopaths and herbalists, including orthomolecular therapists, and some other physicians are now using shitake. Shitake mushrooms contain an antiviral substance known as *lentinan*, which stimulates the immune system to produce more *interferon*, a natural compound that is known to fight viruses. Lentinan is now being studied as a possible treatment for AIDS, and it has been shown to slow the growth of cancerous tumors in animals. Also, shitake mushrooms contain an important type of vitamin D, not found in many foods, but needed for calcium absorption and for the generation of new bone cell growth. Shitake mushrooms also contain *lentysine* and *eritadenine*, two ingredients that help lower the level of fat in the blood, which helps to lower blood pressure and reduce fatigue.

SMOKING CESSATION ⤳ Nicoderm®, Habitrol®, and Nicotrol® are just three of the many products now available to help people stop smoking. The products mentioned here all use a transdermal system, that is, a skin patch, to deliver the active ingredient, nicotine. Persons using these patches while continuing to smoke may experience cardiovascular symptoms, which could indicate heart problems. These products should not be used by pregnant women, individuals with overactive thyroid conditions, or by insulin-dependent diabetics. They should never be used beyond a period of three months under any circumstances.

SODIUM CHLORIDE ⤳ Commonly used as table salt.

SOY ⤳ High in linoleic acid and vitamin E, soy is a good source of protein now being sold widely in a supplemental form. Soy has been shown to contain phyto hormones, among other chemical substances. It is their phyto-estrogens which are beneficial in alleviating the disturbing symptoms most often associated with menopause, such as hot flashes, night sweats, and irritability caused by the highs and lows of estrogen levels. Other studies have shown that soy may have additional redeeming benefits. It has been reported that in certain formulations, soy may actually be able to take up the seating of various cell receptor sites, situated at various locations throughout the body, and which were previously targeted by cancer or HIV/AIDS cells. Studies have indicated that *isoflavones*, naturally occurring substances found in soybeans and tofu, seemed to reduce the rate of mammary cancer by half. Supplements are available as soy protein powder and soy milk.

SPIRULINA ∞ This is a freshwater relative of the seaweed. It is sometimes referred to as blue-green algae. Spirulina is a good vegetable source of vitamin B_{12}. Studies involving spirulina found that inclusion of it in the diet increased the activity of the enzyme *lipoprotein lipasi*, which is involved in fat metabolism. Additionally, a Japanese study on animals detailed that when added to the diet of animals with high cholesterol, spirulina consistently reduced levels of both cholesterol and triglycerides. Wheatgrass, barley grass, and spirulina are excellent sources of beta-carotene, chlorophyll, and minerals. All are helpful for colon disorders, and they are now being recommended by herbalists and naturopaths in their protocols for the HIV/AIDS infected individual, as well as those diagnosed with cancer. Spirulina is available as a powder, in capsules, and as tablets. Pregnant women should avoid spirulina, as it is high in *phenylalanine*, an amino acid.

ST. JOHN'S WORT ∞ *See* Chapter V, Herbs.

STEVIA ∞ This is a relatively new-to-the market supplement. Stevia is now being utilized as a sweetener for replacing refined white sugar in hot and cold beverages. Stevia is more than 100 times sweeter than refined white sugar, so a mere drop or two, when squeezed into a drink, may be sufficient. The newly available powder form lends itself well to some baking recipes. Refined white sugar has been associated with obesity, high blood pressure, coronary heart disease, high serum cholesterol, vitamin B depletion, and more. This sweetener is derived from a plant substance, and it often can be found in combination with chrysanthemum flowers.

Anyone allergic to ragweed will, however, want to avoid products containing chrysanthemum, as it is a relative to the ragweed plant. As of this writing, there are no other known side effects associated with the use of stevia.

SULFUR ∽ Sulfur, in a supplemental form known as *msm* or *methylsulfonylmethane*, is a natural, organic, friendly-to-the-body nonmetallic element. It may become a promising supplement in the near future for encouraging the repair of connective tissues and joint cartilage for those suffering with degenerative bone disorders and inflammatory joint conditions, for wound healing, and for response to allergens.

TAMARIND ∽ Tamarind comes from a large tropical tree, *Tamarindus indica*, native to Africa and the East Indies, and it is preserved in sugar, then used as a natural fruit flavoring. In an extract form it has long been used as a laxative, dating back to early Arab physicians.

TEA ∽ See the medicinal benefits of tea under listing **Green tea** in this chapter.

TEA TREE OIL (Cajuput) ∽ A rendered spicy oil from the white tea tree native to Australia and Southeast Asia, cajuput contains *terpenes, limonene, benzaldehyde, valeraldehyde,* and *dipentene*. This natural disinfectant is commonly used externally for treating fungus infections, athlete's foot, itchy scalp, arthritic pains, as well as an antiseptic for cuts.

THIOTIC ACID ∽ See **Lipoic acid.**

THYME ∽ See Chapter V, Herbs.

TURMERIC ∽ Relative of the ginger family, turmeric is native to East India and most of the Pacific Islands. Indian medicine healers employed turmeric for the treatment of obesity, as it is said to cut the fat from the blood. Presently under study by several universities, including Purdue University, turmeric may soon play a vital role in relieving inflammation, such as the type often associated with arthritis. Just as strong as its anti-inflammatory properties are its anti-bacterial properties. It has shown positive effects in preventing blood clots. Its active principal ingredient, *curcumin*, is also a potent anti-mutagenic substance. We can be certain of hearing a lot more about turmeric and curcumin in the near future.

VITAMINS ∽ *See* separate entries in Chapter II, Nutrients.

VITAMIN A ∽ *See* Chapter II, Nutrients.

VITAMIN B ∽ *See* Chapter II, Nutrients.

VITAMIN C (ascorbic acid, in its natural form) ∽ When taken orally, most of the vitamin is absorbed through the mucous membranes of the mouth, stomach, and upper part of the small intestine. While many experts disagree on a daily recommended intake for this vitamin, most agree that 500 milligrams can be safe and beneficial. The larger the dose, the less is absorbed! For example, ingesting a dose of 250 milligrams results in an 80% absorption rate, while a dose of up to two grams results in about 50% absorption. For this reason, it is best to take vitamin C in small doses several times a day. While considered safe, large doses may cause a slight burning sensation during urination in some individuals, loose bowels, intestinal gas, and for some, skin rashes. Unless taken in the form of sodium ascorbate, large dosages should not be taken by individuals with a tendency toward the formation of kidney stones.

There are no toxic side effects associated with the ingestion of vitamin C, taken in recommended doses. Since it is a water-soluble vitamin, an overdose will likely be excreted by the body via the feces and the urine. While vitamin C is considered relatively non-toxic, overdosing could aggravate an over-acid condition in some individuals due to imbalances caused by poor diets, certain medications, and prolonged states of stress. High doses ingested at one time have produced diarrhea or gas in some individuals, and there is some literature that accuses vitamin C as partly responsible for the development of calcium oxalate kidney stones, kidney disease, and gout. For these reasons, vitamin C is best taken when divided into equal portions at various times of the day.

In my own practice, ascorbic acid is always part of the protocol of recommendations for my HIV/AIDS clients. The RDA for vitamin C is set at 60 milligrams. Nobel Prize winner Linus Pauling and his followers agree that a daily intake of vitamin C should be 2 to 9 grams! To remain on the conservative side, a supplement of 500 to 1,000 milligrams a day should be sufficient. As a supplement, vitamin C is available powdered or as pills and chewable tablets. *See also* Chapter II, Nutrients.

VITAMIN D ∽ *See* Chapter II, Nutrients.

VITAMIN E ∽ *See* Chapter II, Nutrients.

WHEAT GERM ∽ *See* Chapter II, Nutrients, **Vitamin E.**

WHEY ∽ This is the watery part left after the extraction of fat, or curds, from cow's milk. Casein is the principal protein of cow's milk.

YOGURT ∽ Yogurt is a good source of live cultures of lactobacillus acidophilus, as well as a good source of protein. *See* Chapter III, Dietary Sources.

Chapter V: Herbs

In this chapter, we will cover many commonly used herbs, including the most toxic, potentially poisonous, and often misused herbs of the past. These are often still used by those who lack sufficient information regarding their possible toxic effects. But for the most part, we will be discussing the herbs that have recently been received with much enthusiasm by the general public, due to recent mass marketing in the wake of publicized findings regarding their positive effects. However, in order to rule out the manufacturers' hype, we will also be taking a closer look at some prepared herbal supplements, their medical benefits (if any), and how they affect specific disorders. Many of these herbs have gone through chemical analysis and evaluation in order to identify their chemical structures and establish their safety. Others have not! Some of these herbs have been subjected to laboratory studies involving animals.

Most herbs are readily available without a prescription. It is important to know how they interact with specific drugs. For example, individuals taking certain heart or nerve medications should not use certain herbs because they may alter the effectiveness of such medications or cause serious side effects, such as hypotension, hypertension, decreased reactive responses, grogginess, nausea, or heart palpitations. When herbs are utilized for therapeutic purposes, they can often be very effective (for most mild forms of a disorder), but they are not meant to replace the advice and supervision of your medical physician in instances of acute conditions. It is best to seek the advice of a professional (an herbalist or naturopathic doctor) when contemplating the taking of herbs.

In order to effectively use many of the herbs on the market today, keep in mind the following general rules:

(1) When purchasing herbal products or supplements, consumers should only buy those produced by well-respected companies, such as TwinLab. (TwinLab, which has a manufacturing plant easily comparable to any drug manufacturer, employs a full staff of chemists and biochemists.) Others with good reputations include Nature's Way, Nature's Herbs, Puritan's Pride, and Dr. Clayton's. Many of these manufacturers have a board of advisors made up of nutritionists or naturopaths. They often work toward informing the public through newsletters.

(2) When choosing an herbal product, it is strongly recommended that you purchase it from a U.S. manufacturer. Other countries may use inadequate manufacturing processes or have insufficient regulations to monitor product quality.

(3) There are 5 basic delivery systems for taking herbs—pills, capsules, extracts, sprays, and teas. Pills are the least desirable form. Because they are so concentrated and highly condensed, they require huge amounts of water to be metabolized and can often take as many as three days to be utilized by the body. Capsules manufactured with all natural gelatin coverings and containing a fine crystallized essence are good forms of supplementation. An extract of an herb is often preferred, because it is easier to standardize and extracts are often more easily assimilated. Extracts typically contain a carrier of alcohol as a base, and they are

quickly absorbed. An extract (often referred to as a tincture) is made by mixing, steeping, and extracting a particular combination of substances into a desired solution. Today, many manufacturers offer herbs and nutrients as small aerosol sprays. Each spray is measured for potency and is usually provided in a base of purified water, vegetable glycerin, and small amounts of oil. Some sprays may require refrigeration. This type of delivery system is excellent for children and very ill patients. In my practice, I usually recommend extracts first, then capsules. However, for some, drinking an herbal tea or infusion is the safest way to use an herb and the easiest form for the body to metabolize. It is wise to remember that people may build up a tolerance to herbs as some have to antibiotics and other medications due to overuse. To help prevent this, good practitioners usually make their recommendations based on a three week on/one week off regimen, with the ultimate goal of switching their clients to an equally effective herb of somewhat different chemical consistency or weaning their clients off of the herb completely.

We will take a look at the chemical composition and value of many herbs, including some of their contraindications, that is, when it is inadvisable to use them, and their possible benefits. Clients who are seeking the highest degree of positive results must realize that it is important to avoid any antagonists that would inhibit the effectiveness of an herbal supplement.

There are thousands of herbs. In combination with other herbs, there are literally hundreds of thousands of herbal formulas. Like many prescription drugs, herbs often cause side effects when they are taken for prolonged periods of time or in improper dosages. This is why the guidance of a qualified practitioner is emphasized in this book's recommendations. Our individual requirements differ; therefore, it is best to begin with the lowest dosage, gradually making adjustments, until the desired effect is obtained. It is wise to keep in mind that the desired effect may not be noticeable until a few weeks after an individual has begun therapy.

ALFALFA ∞ *Medicago sativa.* *See* Chapter IV, Supplements.

AMARANTH ∞ *Amaranthus hypochondriacus, Amaranthus hybridus L.* Amaranth is a coarse herb, some forms of which grow wild and others which are deliberately cultivated for their seeds, which are ground into a flour. It requires little water for growing and it is believed that amaranth has been around for at least 4,000 years and was originally cultivated in Mexico. **Parts used**: seeds and leaves. **Constituents**: Many species of the amaranth family are now being studied. Its concentrated chemistry composition is not yet available. **Therapeutic benefits**: It has been utilized as an astringent, to relieve diarrhea, to stop bleeding, to treat intestinal ulcers, and to reduce swelling. **Cautions**: none. **Recommendations**: Amaranth flour is available from many health food suppliers. It provides a taste and texture different from the usual bleached white flours. The leaves have been used medicinally in the form of teas or infusions.

ARROWROOT ∞ *Maranta arundinacea.* **Parts used**: roots (rhizomes). **Constituents**: mucilage, starch, and other constituents yet to be discovered. **Therapeutic benefits**: Combined with tepid water, the mashed rhizome resembles corn starch in color and consistency, and similarly, thickens in the same manner. It was once used as a demulcent for relieving inflamed mucous membranes. Arrowroot is more commonly used as a food thickener. At one time, Central and South American Indians relied upon arrowroot for neutralizing wounds made by poisonous arrowheads. **Cautions**: Used in the manners described above, arrowroot is non-toxic.

ASTRAGALUS ∞ *Astragalus membranaceus*. **Parts used**: the root. **Constituents**: glycosides, choline, betaine, rumatakenin, sugar, plant acid, beta-sitosterol, vitamin A. **Therapeutic benefits**: Strengthens the immune system and aids in digestion. Astragalus is one of the most highly regarded herbs used in Chinese medicine. According to the Chinese, it strengthens the body's energy (known as Qi in Chinese terminology). It is also said to support the lungs and has been used in the treatment of debilitating diseases such as HIV/AIDS. The American Cancer Society reported that astragalus root does appear to strengthen the immune system in a large percentage of those taking this herb. It has been part of my own protocol for the HIV/AIDS clients for the past two years, and it has proven to be supportive to the immune systems of these individuals. This herb has also been said to support other glands found throughout the body. **Cautions**: As far as it is known at this point, this herb appears to be completely safe when taken as recommended. **Recommended dosage**: Ten drops of the extract taken in water once daily.

∞ ∞ ∞

BASIL or SWEET BASIL ∽ *Ocimum basilicom*. **Parts used**: leaves. **Constituents**: volatile oils, polyphenolic acids, vitamins A and C, and proteins. **Therapeutic benefits**: It has been shown to be useful for treating some types of skin problems and has also been utilized as an analgesic and an antibacterial for diseases of the kidneys. **Cautions**: Although the oil is said to be non-toxic, one of its volatile oils, estragole, has been proved to cause liver cancers in animals in high dosages. **Recommended dosage**:To create an infusion, place some basil leaves in water and boil or steep until the solution reaches a desired potency. Strain the solution to extract the liquid. The tea has been used for coughs and as a tonic. It is said to be good for stomach problems such as vomiting and gastritis, and for constipation. It may be useful for the common cold in its early stages. Basil was thought to bring about a delayed menstruation.

BLACKBERRY ∽ *Rubus villosus*. **Parts used**: leaves, root and root bark. **Constituents**: tannins, gallic acid, and the leaves also contain flavonoid glycosides of quercetin. **Therapeutic benefits**: Blackberry has been used traditionally as a treatment for diarrhea, an astringent, and to stop bleeding. **Cautions**: Although blackberry is non-toxic, excess doses can trigger constipation and/or diarrhea episodes.

BLACK COHOSH ∽ *Cimicifuga racemosa*. **Parts used**: the dried root and rhizome. **Constituents**: triterpene glycosides (actein and cimigoside), resin (cimicifugin), salicylates, isoferulic acid, tannin, ranunculin (which yields anemonin), and a volatile oil. **Therapeutic benefits**: This herb has been used for various types of pain caused by nerve and muscle disorders.

Black cohosh has been used extensively by Native Americans for treating neuralgia, or nerve pain. Because it is said to depress the central nervous system, it has been employed for treating headache and tinnitus, ringing in the ears. Its resinous compound is said to be responsible for its ability to dilate blood vessels and lower blood pressure, while its salicylates component has made it useful for treating muscular and joint pains, such as those of arthritis. Black cohosh has also been employed as a treatment for asthma due to its antispasmodic abilities. **Cautions**: This is an herb that has powerful chemical properties, and an improper dosage could result in headache, visual disorders, dizziness, nausea and vomiting, and an abnormal pulse rate. **Recommendations**: There are no recommendations offered for this herb. Only a qualified practitioner should supervise the taking of an herb of this type.

BLESSED THISTLE ∽ *See* **Holy thistle.**

BLUE COHOSH ∽ *Caulophyllum thalictroides*. **Parts used**: root and rhizome. **Constituents**: alkaloids, cystine (caulophylline), baptifoline, anagyrine, laburnine, caulosaponin, and resins. **Therapeutic benefits**: This herb was employed initially by Native American women in preparation for childbirth. This herb, not to be confused with black cohosh, might be most useful for treating pain and cramping. **Cautions**: Most experts agree that this herb should not be used during pregnancy, except under the supervision of an expert. **Recommendations**: none. Because there are insufficient clinical studies of this herb as to its effectiveness and safety, I offer no recommendations for blue cohosh.

CALENDULA ∽ *See* **Marigold.**

CAYENNE ∽ **Capsicum frutescens**. *See* Chapter IV, Supplements.

CELANDINE, GREATER ∽ **Chelidonium majus.** Greater celandine is part of a family of herbs known as Papaveraceae. It is often referred to as common celandine or garden celandine. **Parts used**: the entire herb. **Constituents**: It contains the alkaloids caledonian and chelerythrin. Chelerythrin is a narcotic and poisonous. Two of its other alkaloids are homochelidonine A and homochelidonine B. Other chemical substances contained in celandine include protopine and sanquinarine, along with one named doxanthin, a compound considered to be a neutral bitter principle. **Therapeutic benefits**: Celandine has been utilized as an infusion in the treatment of such conditions as jaundice and eczema. In milk, it has been used as an eye lotion to remove white opaque spots on the cornea. In Russia, celandine is still very popularly in use as a treatment (and said to be effective) against cancer.

An infusion includes one ounce of dried celandine mixed with a pint of boiling water. After steeping for several hours, it is then strained to remove the large pieces of herb (leaving only the solution) and stored in a sanitized container, away from heat and/or light. Refrigeration is suggested. Ten drops of this mixed with wine increases celandine's effectiveness in removing obstructions from the liver and gallbladder. **Cautions**: This herb should not be given to children. There seems to be no reported negative effects associated with this herb, but there have not been any studies conducted that confirm its effectiveness and safety. **Recommendations**: Used under the supervision of a qualified practitioner (and in minute amounts), it may be beneficial.

CHAMOMILE ∞ *Anthemis nobilis, Matricaria chamomilla, Chamaemelum nobile L.* **Parts used**: whole herb. **Constituents**: volatile oil, flavonoids, and glucosides. **Therapeutic benefits**: Antispasmodic, antiemetic (stops vomiting), and mildly sedative. Externally, it has been used as an analgesic (reducing pain) for everything from earache to neuralgia. The tea is used to soothe digestion, relieve headache, prevent cramps, and as a mild relaxant. **Caution**: Individuals who have a hypersensitivity to ragweed, asters, and chrysanthemums may also be allergic to chamomile, as it is a species of the same plant family. **Recommendations**: The safest way to take chamomile is as an herbal tea.

CHAPARRAL ∞ *Larrea divaricata, Larrea tridentata.* **Parts used**: the leaves. **Constituents**: The main active constituent of chaparral is nordihydroguaiaretic acid, or NDGA, an antioxidant said to relieve pain and lower blood pressure. Other constituents include amino acids, flavonoids, lignans, volatile oils, and resins. Many herbalists consider chaparral's antioxidant properties valuable for cleansing blood and believe it also contains antibiotic properties. It may be a viable treatment for tumors, cancer, arthritis, flu, and urinary tract infections. Chaparral thrives in mineral-depleted soil and in areas where many plants could not survive. It has been employed in a douche for treating candida. Chaparrals are among the oldest living organisms. It is believed that they have been growing in some areas of California for nearly

12,000 years. Some studies have shown that chapparal has the ability to cause tumor regression. **Cautions**: In 1992, a couple of cases of chaparral-induced hepatitis were reported. **Contraindications**: Due to chaparral's potential toxic effects upon the liver, internal use should be avoided. And certainly, pregnant and lactating women should avoid ingestion of chaparral. **Recommendations**: In my own practice, I've recommended chaparral for external use only. After brewing a commercially available herbal tea made from chaparral, one should allow it to cool, then apply it to inflamed, itchy skin eruptions, such as eczema.

CHASTEBERRY ∽ *Vitex agnus-castus, vitex, chaste tree, hemp tree,* or *monk's pepper.* **Parts used**: the fruit, root, leaves, and flowers. **Constituents**: volatile oil, glycosides, flavonoids, a bitter principle (casstine), and possibly, some alkaloids. **Therapeutic benefits**: Research has shown that chasteberry seems to stimulate the pituitary gland to secrete hormones, which in turn stimulate the corpus luteum to produce progesterone and estrogen. When treating premenstrual syndrome, or PMS, the main goal is to correct the hormonal imbalance, and chasteberry seems to help do this. Chasteberry has been hailed for its ability to alleviate painful breast swelling, thought to be because it contains gamma linoleic acid, or GLA, a precursor of the hormone-like substance prostaglandin. In China, the root, leaves, and fruits are used to prevent malaria and for the treatment of wheezing colds, coughs, and bacterial dysentery. In Indonesia, people use the leaves treat abscesses and ulcers. In Nepal the leaves are smoked to cure headaches, and the juice derived from the leaves is given for the pains of rheumatism. **Cautions**: The dried fruit of the chaste tree contains hormone-like substances that reduce sexual desire in men. **Recommendations**: To ease painful breast swelling, which is associated with PMS, the usual recommended dosage is 3,000 milligrams a day for ten days prior to the expected menstrual period. For other hormonal problems, the recommended dosages may vary. Some herbalists recommend 1/2 teaspoon of a tincture of chasteberry taken before breakfast for several months. Consult a qualified medical practitioner before using this herb.

CHERVIL ∞ *Anthriscus cerefolium*. **Parts used**: leaves, flowers, and seeds. **Constituents**: The chemical constituents of chervil are not fully known. It is a relative of the carrot family. It is not sold in concentrated form and is primarily used as a food seasoning. **Therapeutic benefits**: It has been used in Europe as a brain stimulant for the elderly, and as a diuretic and digestive aid. It has been used externally for gout, abscesses, eczema, and arthritic pain. Because its medicinal powers are associated with its aroma, which is easily lost by heat, it should not be cooked or boiled. French folk healers have proclaimed its usefulness in treating many eye disorders, including detached retinas, glaucoma, and cataracts. **Cautions**: none known.

CLEAVERS ∞ *Galium aparine, clivers.* Cleavers is a species of the family of herbs known as Rubiaceae. **Part used**: the whole herb. **Constituents**: acids; Caffeic, p-coumaric, gallic, p-hydroxy benzoic, salicylic and citric acids. Coumarins are present but unspecified. **Therapeutic benefits**: This herb is said to possess diuretic and mild astringent properties. It is also used for such conditions as psoriasis and enlarged lymph nodes. **Cautions**: This herb is inappropriate for diabetics and pregnant and lactating women, due to the lack of pharmaceutical information at this time. In animal studies, cleavers has been reported to elicit mild laxative effects. **Recommendations**: Under the supervision of a professional practitioner, cleavers, when used as a diluted expressed juice, may assist sedentary individuals in maintaining a healthy lymph system.

CORN SILK ∽ **Zea mays L.** Corn is from the family of Graminaceae herbs. **Parts used**: stigma (the silk) **Constituents**: Amines 0.05% (type not specified); fixed oils (linoleic, oleic, palmitic and stearic acids); saponins 3% (unspecified); tannins up to 11.5 to 13% (unspecified); allantoin, bitter glycosides (1%); cryptoxanthin, cyanogenetic compound (unidentified); flavone, gum, phytosterols (e.g., sitosterol, stigmasterol); pigments, resin, and Vitamins C and K. **Therapeutic benefits**: Corn silk is said to have diuretic and kidney stone-reducing properties. It has been used for conditions such as bed-wetting, bladder infections, prostatitis, and for many chronic inflammatory conditions involving the urinary system. **Cautions**: Allergic responses, such as contact dermatitis and urticaria (hives and rash) have been reported for corn silk. **Contraindications**: Some people have an allergic reaction to corn silk. Excessive dosages may interfere with drug therapy for low blood sugar or for high or low blood pressure. Prolonged use could result in hypokalaemia (abnormally low levels of potassium in the blood) due to its diuretic action. **Recommendations**: Moderation and short term use is wise. Two capsules twice for two to four days, or one to two capsules for an extended period of time (and under a practitioner's supervision) could be considered acceptable. Because the amount of herb in capsules may vary from manufacturer to manufacturer, it is suggested that one consult a physician before using corn silk supplements.

CURRANT ∾ *Ribes nigrum* (black), *Ribes rubrum* (red). *See* Chapter IV, Supplements.

DANDELION ∾ *Taraxacum officinale*. **Parts used**: leaves and root. Dandelion greens are sometimes sold fresh in supermarkets. **Constituents**: acids, minerals (especially potassium), bitter resin, and terpenoids. **Therapeutic benefits**: Traditionally, dandelion has long been recognized for its diuretic actions. Dandelion also has shown an ability to work as a cholagogue (a compound capable of stimulating the flow of bile), and as a laxative. **Cautions**: In low doses, or for short durations, dandelion is considered safe, or non-toxic. However, dandelion may potentiate the action of other diuretics, and could increase any susceptibility toward hypoglycemia. It should be avoided by pregnant or lactating women.

ECHINACEA ∾ *Echinacea angustifolia*, **E. purpurea**, **E. pallida**, Purple coneflower. **Parts used**: dried root and rhizome. **Constituents**: essential oil, humulene and caryophylene; alkaloids; polysaccharide; carbohydrates (including inulin, sugars — glucose, fructose, and pentose); terpenoids;

and glycosides. Other constituents include betaine, fatty acids, phytosterol, resin, and volatile oil. **Therapeutic benefits**: Echinacea contains strong antiseptic, antiviral, and peripheral vasodilator properties, and it is being studied for its potential as an immune system stimulant. In *Herbal Medicines* (1996), Carol A. Newall, Linda A. Anderson, and J. David Phillipson stated that a single 2 ml. subcutaneous injection (stated as equivalent to 0.1 g. of pressed sap), followed by a free interval of one week, stimulated cell-mediated immunity. Daily administration of the injection was stated to have a depressant effect on cell-mediated immunity. The clinical conditions studied have included "infections, arthritis, influenza, colds, upper respiratory tract infections, eczema, psoriasis, urinary tract infections, allergies, gynecological infections, and chronic skin ulcers" (Newall et al., 1996). Echinacea seems to be effective for treating many of these problems. **Contraindications**: There have been no documented contraindications for echinacea; however, it may interfere with immuno-suppressive therapy. For pregnant or lactating women, it might be best to avoid echinacea due to the lack of established safety information. **Recommendations**: A liquid extract (1:1 in 45% alcohol) 0.25 – 1.0 ml. three times a day, or one dropperful mixed in water twice daily for short-term usage.

ELDER, BLACK ∽ *Sambucus nigra* **L** (*Black Elder*) . **Parts used:** flower, leaves. **Constituents**: flavonoids, fatty acids, tannins, glycosides, rutin, quercetin, vitamin C, and many other constituents. **Therapeutic benefits**: Black elder contains anticatarrhal (release of fluid from mucous membranes) and diaphoretic properties (stimulates sweating), diuretic and anti-inflammatory properties. It is an easy to take, fast-acting remedy, beneficial for colds and flus. **Cautions**: Prolonged and/or excessive use can cause potassium depletion. **Recommendations**: Make an infusion of 2 tablespoons of flowers to 1 cup boiling water. Leaves are prepared as a cold remedy; use an infusion made of 1 tablespoon of leaves to 1 cup cold water; let stand 8 to 10 hours.

ELDERBERRY ∞ *See* **Elder**, above.

EPHEDRA ∞ *Ephedra sinica, Ma Huang*. **Parts used**: dried young stems. **Constituents**: alkaloids (such as ephedrine, norephedrine, methyl ephedrine, pseudoephedrine), saponin, flavone, tannins, and essential oil. **Therapeutic benefits**: Ephedra is mostly used in Chinese medicine to treat asthma and hayfever. Its active chemical constituent, ephedrine, is an alkaloid used to relieve nasal congestion, but it is known to constrict certain blood vessels. Ephedrine can also be found in the horsetail plant. Once in common use, ephedrine was found to raise blood pressure markedly, and it is now rarely used to treat asthma. Some herbalists, however, use the whole plant, which contains six other related alkaloids, one of which, pseudoephedrine, may actually reduce the heart rate and lower blood pressure (McIntyre et al., 1988). I for one will not recommend Ma Huang or ephedra to any of my clients. **Cautions**: Individuals with high blood pressure, glaucoma, hyperthyroidism or anyone taking MAO inhibitors or antidepressants should AVOID ephedra! **Recommendations**: none. This herb may have great potential, but as yet its safety and effectiveness have not been proven.

EUCALYPTUS ∽ **Eucalyptus globulus, E. labill.** These are species of the family of herbs known as Myrtaceae. **Parts used**: the leaf. **Constituents**: Flavonoids, eucallyptrin, hyperoside, quercetin, quercitrin, and rutin; volatile oils (including eucalyptol up to 80%) and others of smaller quantities. Eucalyptus also contains tannins, acids, resins, and waxes. **Therapeutic benefits**: Antiseptic, expectorant, and antibacterial properties have been reported, and eucalyptus has shown positive activity against bacterial infections. **Contraindications**: Eucalyptus oil should not be taken internally by pregnant or lactating women, and it should be diluted when used internally or externally. Furthermore, it is said that eucalyptus may interfere with existing therapy for low blood sugar. **Recommendations**: Follow specific guidelines provided by manufacturers on commercial products, and when in doubt, consult a professional.

EVENING PRIMROSE ∽ **Oenothera biennis.** *See* Chapter IV, Supplements.

EYEBRIGHT ∽ **Euphrasia brevipila, E. officinalis L., E. rostkoviana.** Part of the euphrasia species, Scrophulariacea family. **Parts used**: the whole herb. **Constituents**: E. *officinalis* is the commonly used species. It contains acids, alkaloids, amino acids, flavonoids, tannins, and vitamin C. **Therapeutic benefits**: Traditionally it has been employed for sinusitis and, more specifically, for conjunctivitis (when applied locally as an eye lotion). **Contraindications**: The use of eyebright for eye infections has been discouraged. Since the safety of eyebright has not been established due to the lack of pharmacological and toxicity data, it should not be used by pregnant or lactating women. **Toxicity**: Eyebright tincture may induce toxic symptoms. **Recommendations**: The tincture is best avoided. Taken in capsule form under the supervision of your herbologist or naturopath (and used moderately for short durations), it can be beneficial to those who suffer from sinus problems. Two capsules taken daily, with food, at the first sign of a problem should be safe for most, but eyebright should be avoided in a lotion applied to the eye. Consult your practitioner.

FENNEL ∞ *Foeniculum vulgare*. **Parts used**: seed, but the whole herb is used in cooking. **Constituents**: volatile oil, flavonoids, coumarins, fixed oils, and sugars. **Therapeutic benefits:** This herb has long been used for treating colic in infants, as well as abdominal cramps and flatulence. It also has anti-inflammatory properties and is used to stimulate appetite. Hippocrates gave fennel to nursing mothers to stimulate milk production. Fennel is also said to contain antibiotic capabilities, and Pliny recommended it for eye problems, after noting how snakes ate from this plant, immediately after shedding their skins, to restore eyesight. **Cautions**: none known.

FENUGREEK ∞ *Trigonella foenum-graecum* **L.** (Leguminosae family). **Parts used**: seed. **Constituents**: alkaloids, proteins, amino acids, flavonoids, saponins and other constituents such as coumarin, lipids, mucilaginous fiber, vitamins, and minerals. **Therapeutic benefits**: Fenugreek is one of the oldest plants ever used medicinally. Its use can be traced back to the ancient Egyptians. Studies have shown that fenugreek may reduce high blood sugar and may lower high blood pressure as well. When used externally, fenugreek seeds are said to reduce the pain of neuralgia, swollen glands, and tumors. Its saponin-containing plant fibers may inhibit the intestinal absorption of cholesterol. Studies in animals have confirmed that it is able to lower not only cholesterol but also triglyceride levels. Fenugreek has long been used as a laxative and as a nutrient in the treatment of anorexia, dyspepsia, gastritis, and in convalescence. **Contraindications**: Those receiving traditional prescription drug therapy for low blood sugar should avoid fenugreek as it may interfere with the prescribed medications. Caution is also advised for individuals receiving MAOI, hormonal, or anticoagulant therapies in view of fenugreek's amine, steroidal, saponin, and coumarin constituents. **Cautions**: None have been reported when taken in proper doses, except fenugreek is reputed to possibly induce uterine contractions, and thus it should be avoided by pregnant or lactating women. **Recommendations**: For diabetics, fenugreek can assist in the management of insulin, or for those needing to lower cholesterol, two capsules once or twice daily (depending upon practitioner's advice) should be useful.

FEVERFEW ∽ *Tanacetum parthenium*. Part of the Compositae family. **Parts used**: leaf and aerial parts. **Constituents**: volatile oil and tannins. **Therapeutic benefits**: Feverfew is utilized mainly for the relief of headaches (especially migraines) and arthritis. In 1978, several British newspapers covered a story involving a woman who had cured her severe migraine headaches by eating some feverfew leaves each day. In a subsequent clinical study, seven out of ten patients taking feverfew claimed that their migraine attacks were less frequent or less painful, or both. Other clinical studies indicated that the plant has other medicinal benefits, such as alleviating nausea and vomiting, relieving the inflammation and pain of arthritis, promoting sound sleep, and improving digestion. Also, it is said to relieve an asthma attack. Researchers believe that sesquiterpene lactones in the plant may inhibit prostaglandins and histamine released during the inflammatory process, thereby preventing spasms of blood vessels in the head that trigger migraine attacks. **Contraindications**: This herb is contraindicated in people with a known sensitivity to other members of the family Compositae (Asteraceae), which include such plants as chamomile, ragweed, and yarrow. **Cautions**: Its fresh leaves have been said to cause mouth ulcers. **Recommendations**: This herb is valuable when it is known that there are no allergies and when taken in safe recommended doses, as suggested by a professional. In capsule form, the standard dose is usually two capsules as needed. Feverfew is part of a minority group of herbs that have been scientifically evaluated.

FLAX ∞ *Linum usitatissimum*. *See* Chapter IV, Supplements.

GARLIC ∞ *Allium sativum*. *See* Chapter IV, Supplements.

GINGER ∞ *Zingiber officianle*. **Parts used**: the rhizome (root). **Constituents**: volatile oil, mostly zingiberene, phenolic compounds, gingerol, and various other components in smaller quantities. **Therapeutic benefits**: Ginger is an excellent anti-inflammatory, can quiet nausea or vomiting, stimulate appetite, reduce flatulence, and ease stomach spasms. It is reputed to enhance circulation and may help to prevent blood clots. In some studies, ginger was shown to reduce allergic responses, and in most cases, was superior to some types of antihistamines. **Cautions**: Individuals who have insulin problems (such as diabetics or hypoglycemics), heart conditions, or who are taking anticoagulant medications should avoid taking therapeutic doses of ginger. Pregnant or lactating women should avoid therapeutic or continued doses of ginger.

GINKGO ∽ *Ginkgo biloba.* A species which is part of the family known as Ginkgoaceae. Also known as maidenhair tree. **Parts used**: the leaf and seed. **Constituents**: Amino acids, flavonoids, glycosides, and terpenoids are contained in ginkgo's leaves. Its seeds contain alkaloids, amino acids, and phenols. **Therapeutic benefits**: Its leaves and seeds are used in Chinese medicine for lung problems. Recent research indicates that the extract GBE (obtained from yellow autumn leaves) holds a vitamin that improves the integrity of blood vessels and reduces production of tissue damaging free radicals and terpenes, which reduce fibroginin, a clump-forming substance contained in blood platelets. It is also stated that ginkgo increases brain efficiency (thus its now-recognizable ability to improve short-term memory) and cellular energy. Ginkgo is the surviving member of an order of plants dating from 200 million years ago. Ginkgo improves circulation and mental functioning, relieves ringing in the ears, and helps to relieve the symptoms of Alzheimer's disease, coldness and emotional depression, Raynaud's disease, arthritic problems, hardening

of the arteries, dizziness, and anxiety. A synthesized compound is now being studied as a potential drug for preventing rejection of transplanted organs. Ginkgo may have
some efficacy for treating impotence. **Contraindications**:
Ginkgo could cause serious problems for individuals with
clotting disorders. Its fruit pulp has produced severe allergic reactions and should not be handled or ingested. The
seeds can cause severe reactions when ingested. Since there
is insufficient data regarding it effects upon pregnant or lactating women, it is best avoided by them. **Toxicity**: A trial involving more than 100 people treated daily with 120 milligrams of ginkgo biloba leaf extract showed no significant
adverse reactions. At 600 milligrams, there were still no significant adverse reactions to ginkgo biloba. **Recommendations**: Leaf extract—80 to 100 milligrams daily (under a
professional practitioner's supervision). Thirty to 40 milligrams daily is recommended for the seed extract, given
three times daily.

GOLDENSEAL ∞ *Hydrastis canadensis*. *Yellow puccoon*. **Parts
used**: rhizome and root. **Constituents**: alkaloids, volatile
oil, and resin. **Therapeutic benefits**: A strong herb, goldenseal is a tonic for mucus membranes, the liver, and the
uterus. It has been used as a general tonic and eyewash.
Goldenseal is useful as a mouthwash for treating mouth ulcers, and a powder of its rhizome (when inhaled one nostril
at a time) can dilate the sinus passages, making breathing
easier for those who suffer from sinus conditions and allergies. Modern studies confirm goldenseal's potential for
clearing up skin infections, especially impetigo or ringworm,

when used as an external wash. However, it can leave the skin with a lingering yellow tint. Its berberine content makes it a potent antibiotic with antiseptic properties. **Caution**: In 1992, the FDA issued a notice that goldenseal has not been shown to be safe and effective, as claimed in over the counter (OTC) digestive aid products and oral menstrual products. **Contraindications**: Berberine is a known stimulant; therefore, goldenseal should be avoided during pregnancy. **Recommendations**: A liquid extract of 5 to 10 drops taken in water once daily is a safe dosage for most individuals, and is especially useful for repair of irritated mucus membranes.

GOTA KOLA ∽ *Centella asiatica. Gotu kola.* **Parts used**: the whole herb. **Constituents**: Much of this herb's chemistry is unidentified or not readily available. **Therapeutic benefits**: This thick-leaved plant is grown in Pakistan, India, Malaysia, and parts of Eastern Europe. It has been commonly used for treating diseases of the skin, blood, and nervous system. In homeopathy, it is used for psoriasis, blisters, and other skin conditions, and vaginitis. Gota kola has been used in the Far East for treating leprosy and tuberculosis, and as a sedative. Some have written about gota kola's benefits as an immune stimulant, its ability to alleviate the pressure caused to varicose veins, and its ability to accelerate the healing of burns. **Recommendations**: Due to the present lack of information regarding this herb, including its chemical constituents, it is not recommended for use. However, studies are ongoing, and new information will soon be made available. Gota kola should not be confused with kola nut, the caffeine-containing nut used in soft drinks.

HAWTHORN ∞ *Crataegus oxyacantha*. **Parts used**: flowers, fruits. **Constituents**: Hawthorn belongs to the family Rosaceae. This herb contains alkaloids, is bitter to the taste, and is soluble in water. It is a spring-flowering thorny shrub containing flavonoids (including quercetin) and the flavone rutin. Other constituents include tannins, glycosides, and saponins. **Therapeutic benefits**: Native to Europe, North America, western Asia, and parts of Europe, hawthorn can be traced back to the time of Christ and was thought to have furnished the crown of thorns. The hawthornberry is said to possess properties that are very helpful to the heart, and it has been used traditionally for cardiac failure, myocardial debility, hypertension, and arteriosclerosis. When hawthornberry extract was given to laboratory mice, it was reported to enhance enzyme action. Studies on human subjects showed overall improvement of cardiac function and of subjective symptoms. **Contraindications**: Nausea and fatigue, sweating, and a rash on the hands have been reported as side effects in clinical trials using commercial preparations of hawthorn. Because hawthorn has been reported to exhibit many cardiovascular activities, it may affect the existing therapy of patients with various cardiovascular disorders, such as hypertension and hypotension. **Toxicity**: Acute toxicity of isolated constituents (mainly flavonoids) has been documented. **Recommendations**: It is difficult to provide an across the board dosage that will be compatible with all individuals. Certain specific considerations need to be addressed, including age, diagnosis, and other prescribed medications that may be part of a treatment plan. A capsule of 450 milligrams (taken 3 at a time) twice to three times a day is the usual recommendation for high blood pressure and several other heart ailments. This herb should only be taken with a physician's knowledge, and only under supervision of a qualified professional.

HOLY THISTLE or BLESSED THISTLE or MILK THISTLE ∞ *Cnicus benedictus*. **Parts used**: the whole herb. **Constituents**: lignans, polyenes, steroids, tannins, terpenoids, volatile oils, mucilage, nicotinic acid, and resin. **Therapeutic benefits:** This annual herb is native to southern Europe, is cultivated in gardens in many parts of the world, and has been used medicinally since A.D. 100. It has been utilized for relieving excess fluid to inducing sweating, as an herbal to induce menstruation, to help treat indigestion, and to stimulate appetite. Because of thistle's high lignin content, it is used today by herbalists to combat replication of the virus responsible for HIV. It has also been found to provide some positive effects when used to treat hepatitis and other liver disorders. It is said to possess antidiarrheal, antihemorrhagic, expectorant, antibiotic, and antiseptic properties. Also, there is some documentation of antitumor activity. **Toxicity**: None has been documented. **Contraindications**: None have been documented. However, holy thistle may cause an allergic reaction in individuals with a known hypersensitivity to other members of the Compositae family, which include chamomile, ragweed, and tansy. Because the safety of holy thistle has not yet been established, it is best avoided by pregnant and lactating women. **Recommendations**: 1.5 to 3.0 milligrams of a liquid extract to be taken according to suggestions or recommendations of a naturopath or herbalist.

HOPS ∞ *Humulus lupulus, Silent night.* The hop is a vine of the hemp family. (Cannabaceae). The dried ripe cones of the female flowers are used in the making of beer and ale and account for their bitter taste. **Parts used**: the dried female flowers. **Constituents**: volatile oil, resin, acids, and over 100 other compounds, including tannins, glycosides, flavonoids, fats, amino acids, and estrogenic substances. **Therapeutic benefits**: This widely cultivated plant has been used in folk medicine for its calming effect on the body. It has also been used to relieve cramps, stimulate the appetite, as a poultice to relieve sciatica, arthritis, toothache, and other nerve pain, and a tonic to induce sleep. **Contraindications**: Allergic reactions have been associated with hops, although only after external contact was made with the herb and oil. It is thought that people who suffer from depression or who are taking medications for the same should not take hops, as the herb might reduce the effects of existing therapy. It is advisable that pregnant or lactating women avoid hops. Due to its estrogen-like substances, it has been noted that men who consume excessive amounts of beer (which largely contains hops) may begin to note an increase in the size of their breasts. **Toxicity**: Handling hops cones is not recommended. Respiratory allergy has been documented. **Recommendations**: 0.5 to 2.0 milligrams of the liquid extract of hops has been the standard recommendations of many herbalists. As with any of the herbs, it is wise to consult a qualified practitioner. *See also* Chapter IV, Supplements.

∞ ∞ ∞

HORSERADISH ∽ *Armoracia lapathifolia, Radicula armoracia L.* **Parts used**: the root. **Constituents**: coumarins, phenols, volatile oil, enzymes, starch, and sugar. **Therapeutic benefits**: It is a circulatory stimulant, has mildly antiseptic properties, is a digestive aid, and a diuretic. It is excellent for asthmatic, bronchial, and lung disorders. **Cautions**: Individuals who have been diagnosed with a hypothyroid condition or individuals taking the drug thyroxine should not take this herb. This fact is said to extend to other members of the cabbage and mustard family.

HORSETAIL ∽ *Equisetum arvense. Shave grass.* **Parts used**: aerial parts. **Constituents**: Almost three-quarters of horsetail's chemical substance is made up of silica. The remaining 25% of its chemical structure is saponins, small bits of tannins, alkaloids, flavonoids, as well as the minerals manganese, sulfur, magnesium and potassium. **Therapeutic benefits**: Its high content of silica makes this herb useful for those suffering from certain bone disorders (including osteoporosis). Silica is an essential mineral, along with other specific minerals such as calcium, boron, and copper. Silica has been shown to be useful in the regeneration of weak, splitting hair and nails. Horsetail has been used medicinally for urinary tract infections, arteriosclerosis, and lung disorders. It also has mild diuretic properties and was once used to stimulate urination. **Cautions**: Horsetail is safe when used properly and under the guidance of a professional. **Recommendations**: Under ordinary circumstances, two capsules taken daily with water is considered to be an adequate supplement. For therapeutic goals and dosages, consult your practitioner.

HYSSOP ∽ *Hyssopus officinalis* is part of a family known as Labiatae. **Parts used**: the leaves and the flowering buds. **Constituents**: volatile oil, flavonoids, and tannins. **Therapeutic benefits**: This herb is ancient and can be found mentioned several times in the Bible. It was once used by Hippocrates for pleurisy. It is said to aid in the digestion of fatty meat, and it was once used for purifying temples. The leaves contain antiseptic, antiviral oil. A mold that produces penicillin grows on its leaves. It has been employed as a mild sedative and an expectorant for influenza and bronchitis, and it has been said to be a blood purifier. It is an herb that I have used in my practice to help support the immune systems of individuals with HIV/AIDS who for various reasons cannot tolerate the side effects associated with traditional drug therapy. **Cautions**: There is not enough information available yet regarding the safety of hyssop, but there have been no reported cases of toxicity associated with its use, when used in proper dosages and under the supervision of a qualified practitioner. As in the case of many potent herbs, it is recommended that pregnant women or lactating mothers avoid using hyssop. **Recommendations**: Most practitioners recommend an extract of the leaves. Eight to 10 drops in water daily (short-term use, or an on-again, off-again schedule formulated by one's qualified practitioner).

IRISH MOSS ∽ *Chondrus crispus* (dried red algae). **Parts used**: whole herb. **Constituents**: This herb is an nutrient and dietetic with soothing properties. It contains many vitamins as well as chlorides, iodides, bromides, phosphates, sulfates of potassium, calcium, and magnesium. Many biochemists refer to Irish moss as the chemistry of life. It contains five polysaccharide complexes known as carageenans, which make up about 80% of its chemical composition. **Therapeutic benefits**: Irish moss is a valuable nutrient for the week and feeble, therefore, it is often recommended for those who suffer debilitating diseases. Because Irish moss has a good amount of sodium sulfate (each single molecule of sodium sulfate has the ability of taking up and carrying out two molecules of water), it is often an effective remedy for edema (water retention). For those deficient in certain minerals, Irish moss is, once again, invaluable. Its mucilaginous consistency lends itself to the treatment of chronic lung and upper respiratory problems. Herbalists have long known the benefit of utilizing Irish moss for the treatment of bronchitis. **Contraindications**: None have been documented or recorded for Irish moss. **Recommendations**: Look for Irish moss in its dried form and combine with a food for eating, such as in almond butter, or drink as herbal tea.

LADY'S SLIPPER ∽ *Cypripedium pubescens*, related to the Orchidaceae family. **Parts used**: root, rhizome. **Constituents**: While a thorough investigation has not yet been completed regarding the chemistry of lady's slipper, several constituents have been identified. They include glycosides, tannic acid, resin, gallic acids, volatile oil, and tannins. **Therapeutic benefits**: Lady's slipper has been used in herbal extract forms because of its sedative and antispasmodic abilities. It has also been used for treating hysteria, nervousness, anxiety, and insomnia. **Caution**: It has been said that the roots may cause delusions or hallucinations. Larger doses could result in headache, a state of heightened excitability, and visual delusions. **Contraindications**: Lady's slipper may cause allergic reaction in sensitive individuals. This herb should be avoided by pregnant or lactating women. **Recommendations**: Several variations of this herb are available in extract form, but should only be recommended and supervised by an herbalist or doctor of naturopathy.

LICORICE ∾ *Glycyrrhiza glabra* **L. Parts used:** the root. **Constituents**: coumarins, flavonoids (including isoflavones), amino acids, and other traces of chemical substances. **Therapeutic benefits**: Licorice works well as an expectorant and was once used for the treatment of ulcers. Reportedly, it was used with great success in China for treating hepatitis B. Under strict supervision, licorice root has been employed with reported but undocumented success for the treatment of chronic fatigue syndrome. **Cautions**: The phytochemistry is well documented for licorice, and it is, therefore, known to effect hypertension, or high blood pressure, in a number of individuals who have used this herb. It should be avoided by individuals with an existing cardiovascular-related disorders. Due to this herb's estrogenic and steroid effects, it is suggested that pregnant or lactating women avoid it. It may also interfere with hypoglycemic therapy. **Recommendations**: In view of the documented information surrounding licorice, no recommendations are offered.

MA HUANG ∾ *See* **Ephedra**.

∾ ∾ ∾

MARIGOLD ∽ *Calendula officinalis*. **Parts used**: flowers.
Constituents: carotenoids, essential oil, flavonoids, and
other chemical compounds such as saponins and mucilage.
Marigolds also contain fat, sugar, potassium chloride, potassium sulfate, calcium sulfate, and sodium. **Therapeutic
benefits**: A salve containing cut walnut leaves, echinacea,
eucalyptus, and marigold flowers has been found to be useful for treating swelling, sprains, bruises, and skin cancers.
In combination with the herb myrrh, marigold has long been
recommended by herbalists for treating some types of
rheumatism. Marigold has the ability to stimulate the flow
of bile, thus it is useful as a digestive aid and is a good remedy for ulcers, including duodenal ulcers. Marigold has long
been available in commercial preparations (lotions and
creams) for alleviating certain types of skin conditions, including eczema. Herbalists have long employed the benefits of marigold for halting bleeding, and it has been reported to lower blood pressure. **Caution**: probably best
avoided by pregnant women. **Contraindications**: Marigold
flowers are thought to be safe. No problems or side effects
have been found in connection with the use of marigold.
As far as external use is concerned, it is a very soothing and
healing remedy. **Recommendations**: For internal use, seek
the advice of a professional for specific recommendations.
Externally, it is absolutely safe.

MARSH MALLOW ∽ *Althaea officinalis*. A species belonging
to the family Malvaceae. **Parts used**: the root and the leaf.
Constituents: The main active ingredient of marsh mallow
appears to be its mucilage, which is about 30% of total com-

position. Other constituents include asparagine, calcium oxlate, tannin, pectin, phenolic acids, and starch. **Therapeutic benefits**: Marsh mallow root has been employed as a mouthwash and gargle, or for soothing the pain of teething for infants. Its high mucilage content makes marsh mallow an ideal remedy for bronchitis and asthma. Marsh mallow is useful as an expectorant for tight coughs. The pulverized root can be utilized externally as a drawing poultice (applied warm). **Contraindications**: none have been recorded. **Recommendations**: An extract given (5 to ten drops) in water twice daily will alleviate a cough and sooth the mucous membranes.

MILK THISTLE ∞ *See* **Holy thistle.**

MINT ∞ **Mentha species.** *See* Chapter IV, Supplements. *See* **Peppermint,** below.

MOTHERWORT ∞ *Leonurus cardiaca* **L.** Part of the Labiatae family. **Parts used**: the whole herb. **Constituents**: alkaloids, flavonoids, tannins, and terpenoids. Other chemical constituents include carbohydrates, choline, and other compounds. **Therapeutic benefits**: There are several species of this family of Labiatae herbs, but not all have been chemically studied. *Leonurus artemisia* is commonly used in Chinese medicine. Other species referred to as motherwort include L. *sibirious* and L. *heterophyllus*. But it is the species of L. *cardiaca* (the European species) for which the chemistry has been well studied. Cardioactive properties of mints have been studied in animals, and they support the use of this species of the herb. It has been employed for its sedative and antispasmodic abilities; therefore, it has been used in some cases for cardiac debility and simple tachycardia, or rapid heart-beat. **Contraindications**: Excessive use and/or improper dosages may interfere with existing therapy for cardiac disorders. Sensitive individuals may experience an allergic reaction. In view of the lack of information regarding toxicity data, it is suggested that motherwort be avoided by pregnant and lactating women. **Side effects**: The leaves of this herb may cause contact dermatitis, and the lemon-scented oil may result in photosensitization. **Recommendations**: This herb could be useful to some individuals who suffer from spastic colon, menstrual cramps, and other similar problems. It is my recommendation, however, that this herb be used under the specific guidelines of an herbalist or other qualified practitioner, as dosage and duration will vary from one individual to another.

MULLEIN ∽ **Verbascum thapsus** is a species belonging to the family known as the Scrophulariaceae. **Parts used**: flowers and leaves. **Constituents**: mostly mucilage. Its other compounds include saponins and flavonoids. **Therapeutic benefits**: Here is an herb that (due to its high mucilage content) is a soothing expectorant and is, therefore, recommended for coughs. Research confirms mullein's antitubercular activity in plant extracts. Smoke from the burned leaves were used by Native Americans to revive the unconscious. The flowers reduce eczema inflammation and help to heal wounds. The root of this herb is diuretic, and, in homeopathic medicine, a tincture of the leaf is used for treating migraine headaches and earache. This is an herb that I often suggest for short-term use, for individuals who complain of minor lung problems. **Contraindications**: None were available for mullein. Based on all the information that I could gather on this herb, I have concluded that it is apparently non-toxic. **Recommendations**: For anyone who is under his/her physician's care for a serious medical condition, consult with your doctor prior to using this or any other of the herbs mentioned herein. Six to ten drops of a liquid extract (considerations for dosage include age, weight, etc.) of the leaves used once daily (in the A.M.) for two to three days should be useful.

OATS ∞ **Avena sativa.** *See* Chapter IV, Supplements, *Oat fiber.*

OLIVE ∞ **Olea europaea.** *See* Chapter IV, Supplements, *Olive oil.*

OREGON GRAPESEED ∞ *See* Chapter IV, Supplements, *Grapeseed.*

PAPAYA ∞ **Carica papaya**. *See* Chapter IV, Supplements.

PASSIONFLOWER ∞ **Passiflora incarnata** belongs to the family known as Passifloraceae. **Parts used**: the entire herb. **Constituents**: alkaloids, flavonoids, several acids, including the essential fatty acids, linoleic and linolenic acids. Other constituents include several other acids, sugars, and gum. Found in the root of this plant are other compounds, especially coumarins. **Therapeutic benefits**: Passionflower is said to possess sedative, hypnotic, antispasmodic, and anodyne properties. It has been used traditionally for neuralgia, generalized seizures, hysteria, nervous tachycardia (rapid heartbeat), spasmodic asthma, and specifically for insomnia.

It has also been said to contain a compound known as serotonin, and since serotonin is often deficient in individuals who suffer from mild forms of depression, passionflower is often recommended by herbalists (in combination with other herbs such as valerian, skullcap, etc.) for alleviating symptoms of mild depression by compensating for serotonin deficiencies. It is an herb that I would be comfortable suggesting to any of my clients, provided, of course, they are not being medically treated for depression. In such instances, herbal recommendations might interfere with medical treatments by negatively interacting with the prescribed treatment. Individuals who suffer from serious diagnosed medical conditions and who are under the treatment of a medical physician should consult with him or her prior to embarking on any self-treatment plan. **Contraindications**: Excessive dosages of passionflower may cause sedation. Both harman and harmaline have exhibited uterine stimulant activity in animal studies (and both are present in passionflower). In view of these data, passionflower should only be used by pregnant or lactating women in mild doses and only then, under the guidance of a qualified professional. **Recommendations**: Seven to ten drops of a liquid extract mixed in water and taken once or twice daily, depending upon one's symptoms, severity of same, and a qualified practitioner's suggestions. And, as always, considerations such as age, weight, and other factors must be evaluated. No reported side effects of passionflower use were found. Passionflower in combination with valerian root is effective for inducing sleep, calming nerves, and is said to strengthen the heart.

PAU D' ARCO ∞ *Tabecuia heptaphylla* or *impetiginosa*. *Lapacho*. **Parts used**: the inner bark. **Constituents**: The chemistry of this herb is not recorded and not much information is available at this writing. **Therapeutic benefits**: What is known and agreed upon by most individuals who study the chemical constituents of herbs (including many of the biochemists who are retained by the various manufacturers of herbal products) is that pau d' arco has antibiotic and antiparasitic properties. It is said to be an anti-cancer herb that strengthens the immune system. It is also stated that this herb lowers blood sugar, and is now being studied in Brazil as a treatment for cancers, including leukemia. **Recommendations**: For any serious illness, always consult a qualified practitioner. For other temporary uses, low dosages that adhere to the manufacturer's suggested recommendations are best (potencies may vary). It is not recommended for pregnant or lactating women.

PEPPERMINT ∞ *Mentha piperita*, **M.** *arvensis*, **and M.** *spicata* all belong to the mint family, Labiatae. **Parts used**: fresh and dried aerial parts, leaves. **Constituents**: volatile oil, primarily menthol, tannins, flavonoids, tocopherols, choline, and a bitter principle have been documented and recorded as part of the chemistry constituents contained within the various species of the mint family. **Therapeutic benefits**: Peppermint (and the oil made from the dried leaves) is native to European, Asian, and American gardens. The Chinese have long employed peppermint as a remedy for head colds and influenza. It is utilized as a flavoring in many food products, and it has been used in the treatment of certain types of headache, sore throats, and, also, in a solution for eye inflammations. Herbalists say it is a muscle relaxant and aids digestion. In early medicine, it was employed for spasm of the bowel. As a digestive aid, it is said to stimulate the flow of bile. Recent studies have found peppermint to contain anti-ulcer, anti-inflammatory, and liver bile-stimulating properties. Peppermint is further said to inhibit the growth of many kinds of germs. **Caution**: Peppermint has been shown to cause allergic reactions such as hay fever and skin rash in sensitive individuals. Avoid prolonged use of the essential oil as an inhalant. Mint can irritate the mucous membranes and should not be given to children for more than a week without a pause. In 1992, the FDA issued a notice that peppermint and peppermint spirit have not been shown to be safe and effective as claimed in OTC (over-the-counter) digestive aid products, insect bite and sting drug products, oral menstrual drug products, and in astringent drugs. **Contraindications**: None have been recorded. **Recommendations**: Peppermint oil should not be taken internally by children, or by pregnant or lactating women. All other individuals should consider diluted forms (including commercially packaged products, according to label directions), or seek the council of an authority. *See* Chapter IV, Supplements, for more information.

POKEROOT or POKEWEED ∽ *Phytolacca americana* **L.** is a species belonging to the Phytolaccaceae family. **Parts used**: root. **Constituents**: alkaloids, lectins, saponins, some acid compounds, including histamine and GABA. **Therapeutic uses**: Pokeweed, pocan, or red plant is stated to possess antirheumatic and mild pain-relieving properties. It has anti-parasite and anti-fungal properties as well. Traditional uses were for rheumatism, inflammation of a mucous membrane of the nose or throat, tonsillitis, laryngitis, and skin infections. **Contraindications**: Fresh pokeroot is poisonous and the dried root emetic and cathartic. The toxic effects documented following ingestion of pokeroot make it unsuitable for internal ingestion. In addition, external contact with the berry juice should be avoided; systemic symptoms of toxicity have occurred following exposure of broken skin and conjunctival membranes to the juice. In 1979, the American Herb Trade Association declared pokeroot should no longer be sold as an herbal beverage or food. **Recommendations**: Avoid!

RED ELDER ∾ Red elder is toxic!

ROSEMARY ∾ *Rosmarinus officinalis* L. **Parts used**: leaf, twig. **Constituents**: flavonoids, phenols, volatile oil, and terpenoids. **Therapeutic benefits**: Rosemary has anti-microbial and anti-parasitic properties and is considered to be a stimulant, especially of bile flow. It has been utilized to halt spasms, as a mild sedative, and as a diuretic. Traditionally, rosemary was used for headaches, indigestion, and topically in an ointment for myalgia. **Cautions**: Because rosemary contains as much as 20% camphor, it can cause convulsions if taken in sufficient doses. In topical preparations such as cosmetics, or bath preparations, rosemary oil can cause skin problems in people who are hypersensitive. It should be avoided by pregnant or lactating women.

PUMPKIN ∾ *Cucurbita pepo.* See Chapter IV, Supplements, Pumpkin seed.

SAFFLOWER ∾ *Carthamus tinctorius.* See Chapter IV, Supplements.

SAINT JOHN'S WORT ∾ *See* **St. John's Wort**.

SARSAPARILLA ∾ *Smilax* species, S. *aristolochiifolia* Mill, S. *regelii* Killip et Morton, S. *officinalis* Kunth and S. *febrifuga*, all belonging to the family Liliaceae. **Parts used**: root rhizome. **Constituents**: saponins, acids, quercetin, resin, starch, volatile oil (minute amount) and cetyl alcohol. **Therapeutic benefits**: Laboratory studies show sarsaparilla to contain anti-inflammatory and liver-protective functions. Human studies suggest that this herb stimulates appetite, aids digestion, and is useful as a diuretic. Limited clinical data utilizing extracts indicate improvement in psoriasis. Pharmaceutical comments are based on sarsaparilla's steroidal saponin constituents. Available scientific evidence fell short of justifying herbal uses. No toxicity data were located, although large dosages may be irritating to the gastrointestinal mucosa and should, therefore, be avoided. Sarsaparilla saponins have been used in the partial synthesis of cortisone and other steroids. Sarsaparilla has long been recommended by many herbalists for balancing hormonal activity. While it does not contain any detectable active hormones, it is said to have the ability of stimulating the production of certain hormones, when required, and/or inhibiting the production of other hormones when they are overly abundant. **Contraindications:** None have been documented for sarsaparilla however, in view of the irritant nature of its saponin constituents, excessive ingestion should be avoided. There are no known problems with the use of sarsaparilla during pregnancy and lactation; yet, in light of the possible irritant nature of its saponin components, again, excessive ingestion should be avoided. **Recommendations**: A liquid extract of its dried root (1 to 4 grams) or ten drops taken in water, up to three times a day, have been suggested by herbalists, depending upon such variables as desired effect, age, and weight. Always seek the advice of a qualified practitioner.

SAW PALMETTO ∾ **Serenoa serrulata** is a species of the Arecaceae/Aalmae family. **Parts used**: the fruit. **Constituents**: carbohydrates, fixed oils, steroids, including trace amounts of unidentified compounds. **Therapeutic benefits**: Colds, mucous inflammations, reproductive disorders, and urinary diseases have all been treated with saw palmetto. It has been used for treating prostatitis, or inflammation of the prostrate. Its effectiveness is probably due to its steroidal saponin content. In some parts of Europe, saw palmetto is used in over-the-counter remedies for benign prostate enlargement. North American Indians discovered that the berries were nutritive for humans and had a soothing effect upon mucous membranes throughout the body. **Contraindications**: Since there are insufficient data regarding the safety of using saw palmetto, it is not recommended for pregnant or lactating women. Due to this herb's estrogenic activity, it could affect existent hormonal therapy, including the oral contraceptive pill and hormone replacement therapy.

Recommendations: Liquid extract of the fruit; 0.5 g (one dropperful) in water, up to three times a day. Discuss proper recommendations with your practitioner. *See also* Chapter IV, Supplements.

SELF-HEAL ∞ ***Prunella vulgaris*. Parts used**: aerial parts. **Constituents**: bitter principles, volatile oil, and alkaloids. **Therapeutic benefits**: Self-heal has been used as a gargle for sore throats and in a mouthwash. **Contraindications**: None are documented. **Recommendations**: Use as directed at the first sign of a sore throat.

SKULLCAP ∞ ***Scutellaria lateriflora* L. and S. *baicalensis georgi*** belong to the Labiatae family. **Parts used**: entire herb. **Constituents**: There has not yet been a complete analysis of the chemical compounds contained in skullcap; however, flavonoids and volatile oils have been identified. **Therapeutic benefits**: Skullcap has been shown to possess sedative and antispasmodic abilities. Presently, it is employed by many herbalists in the treatment of epilepsy, convulsions, neuralgia, insomnia, tetanus, and mild forms of anxiety. Some scientific studies prove that skullcap stabilizes blood pressure. Skullcap is also used as a mild tranquilizer. **Caution**: Liver toxicity cases have been reported to the Centers for Disease Control caused by the use of skullcap. **Toxicity effects**: Giddiness, confusion, and seizures have been caused by overdoses of scullcap tincture. **Recommendations**: In view of the liver toxicity associated with skullcap, it is best to avoid it. Therefore, no recommendations are offered. Individuals who insist on using skullcap in spite of the warnings listed, should consider using it only in diluted forms, such as in a commercial herbal tea.

SLIPPERY ELM ∽ *Ulmus rubra,* a species that is part of the Ulmaceae family. **Parts used**: the inner bark. **Constituents**: carbohydrates, with mucilage being the major constituent. Other compounds include tannins and calcium oxalate. **Therapeutic benefits**: Slippery elm is a nutrient, and as a nutrient it is very well tolerated and easily assimilated by the weak and feeble. It is a soothing and healing herb to all of the body's mucous membranes, due mainly to its high mucilage content. It has been utilized for conditions associated with inflammation of the digestive tract or the lower bowel, and it is useful in alleviating the symptoms of diarrhea or colitis. Topically, it is beneficial when used as a poultice for treating abscesses or boils. Because of slippery elm's high mucilage content, it is also beneficial in assisting individuals with high blood lipids to lower their blood cholesterol levels. This herb has been used for many years by herbalists in the treatment of cancer (in combination with several other herbs). It is able to soothe and heal tissues and to assist in the regeneration of new tissue. **Contraindications**: Slippery elm is safe, and there are no contraindications associated with the use of this nutritive herb. **Recommendations**: Because of this herb's soothing properties and non-toxic status, it is, more often than not, recommended in combination with other herbs for affecting various specific responses.

∽ ∽ ∽

SOAPWORT ∾ *Saponaria officinalis*, a species that is part of the Caryophyllaceae family. **Parts used**: rhizome. **Constituents**: the active principle is its saponin content. **Therapeutic benefits**: This herb was given its name because it produces lather when agitated in water, and its active ingredient, saponin, is a detergent. Its leaves provide an extract that has been used to promote sweating, as a remedy against rheumatism, and to purify the blood. **Contraindications**: In large amounts soapwort is a strong purgative, and mildly poisonous. **Recommendations**: None offered. This is an herb that should be recommended and supervised by a qualified professional.

STAR ANISE ∾ *Illicium anisatum, Illicium verum*. **Parts used**: fruit. **Constituents**: volatile oil, fixed oil, choline, mucilage, and sugar. **Therapeutic benefits**: A carminative (relieves gastrointestinal flatulence) and an expectorant often used in cough preparations. It has some diuretic effect and anti-bacterial qualities, but may cause skin reactions in sensitive individuals. **Cautions**: As a food it is non-toxic. In China it is also used as an aid for rheumatism, and as an ingredient in recipes, especially in desserts.

STINGING NETTLE ∾ *Urtica dioica* is a species that is part of the family Urticaceae. **Parts used**: the aerial parts of young plants. **Constituents**: formic acid, acetylcholine, chlorophyll, and several vitamins and minerals, such as iron, potassium, manganese, silica, sulfur, and vitamins A and C. **Therapeutic benefits**: This herb has long been utilized in shampoo and rinses for the hair, and in some foods, especially soups. Stinging nettle (or nettle), is an old fashioned remedy for backache and is excellent for eczema. More recently, in laboratory studies, nettle has been shown to possess anti-inflammatory abilities and is able to lower sugar levels in the blood. Some herbalists have recommended nettle to those individuals who are prone to or presently do have mild degrees of anemia. **Caution**: The uncooked plants can cause kidney damage and poisoning. **Contraindications**: There are no available data regarding reported toxicity associated with the use of nettle. **Recommendations**: An herbal tea made from nettle is recommended.

ST. JOHN'S WORT ∞ *Hypericum perforatum*. *God's wonder herb*, *Grace of God*, **and** *goatweed* are other names this herb has been given. More recently, it has been called "nature's Prozac®." **Parts used**: aerial parts. **Constituents**: glycosides, and in particular a red pigment, hypericin. Other compounds include flavonoids, phenols, tannins, resin, and volatile oils. In pharmacological studies, the active principle of St. John's Wort was said to be of the quercetin type. **Therapeutic benefits**: This is an herb that has a variety of multiple benefits, and it has been employed extensively in herbal and homeopathic remedies, including the treatment of neuralgia. St. John's Wort contains sedative and astringent properties; thus, it is widely used today by herbalists and lay people for treating mild forms of depression. Also, it has been found to be effective for assisting the HIV/AIDS diagnosed individuals in maintaining a healthier immune response. St. John's Wort is also said to possess diuretic properties, and is very useful when applied externally for easing the pain of neuralgia. It is also useful for calming the nervous system.

Cautions: A photosensitivity has been documented for St. John's Wort, as well as an allergic response in some individuals causing skin eruptions; however, these reactions were documented after ingestion of an herbal tea made from the leaves. **Contraindications**: Excessive dosages may have an adverse reaction with prescribed drugs. It may cause allergic reaction in sensitive individuals (hypericin). Because of insufficient data regarding St. John's Wort's safety for pregnant or lactating women, it is best avoided by such individuals. **Caution**: This herb should not be introduced to any individual who is already on any medication for the treatment of depression. To do so could potentially cause serious side effects. Those individuals who wish to make a switch from a prescribed medication for treating depression to an herbal remedy such as St. John's Wort should first speak to their physician. With their doctor's guidance (and in conjunction with a qualified herbalist or naturopath) the patient can begin to gradually be weaned from his/her present medication, and, after a proper amount of time has elapsed, St. John's Wort could be recommended at the proper potency and frequency. **Recommendations**: A liquid extract of the herb taken in water (ten drops from a dropper), once daily is the usual recommendation. One should consult an herbalist or other qualified practitioner regarding stronger dosages.

TANSY ∽ *Tanacetum vulgare* **L.,** a species belonging to the family of Asteraceae/Compositae. **Parts used**: the whole herb. **Constituents:** steroids, terpenoids, volatile oils and at least ten different chemotypes have been identified in which camphor was the most frequently occurring main component and thujone was second. **Therapeutic benefits**: Tansy is said to contain antispasmodic abilities and is a stimulant to abdominal viscera (internal organs). **Caution**: Tansy oil contains ketone-thujone, which is toxic. Symptoms of tansy oil poisoning (rapid or weak pulse, violent spasms and/or convulsions) are attributable to the thujone content. **Contraindications**: Tansy oil is toxic! Deaths have been reported following internal ingestion of infusions and extracts. **Recommendations**: None. There are insufficient studies as of this writing to justify using this herb, especially in the hands of the inexperienced. This is not an herb that should be taken as part of a self-treatment plan. Tansy has undergone several laboratory studies and, in diluted laboratory formulas and dosages, there were some positive results, yet not enough to warrant its use. Perhaps one day in the near future, a diluted, safe, and tested formula will surface.

THYME ∞ *Thymus vulgaris*. **Parts used**: the whole herb. **Constituents**: Distillation of the leaves and flowering tops produce thyme oil, which is sold commercially. The main constituents are thymol and carvacrol, which are phenols. Volatile oils are present, as are tannins and several acids, including some flavonoid content. **Therapeutic benefits**: Antitussive (suppresses coughing), antiseptic, expectorant, and antispasmodic. **Cautions**: Can be toxic in overdoses and should be used with care! Pregnant and lactating women should avoid the use of thyme oil. **Recommendations**: Thyme is safest when used properly as an ingredient in a regulated remedy preparation. It is used in diluted form as a tea. Thyme tea can also be used as a gargle for sore throats and a drink for the common cold, influenza, fever, and allergies. Make the tea by steeping a dozen fresh sprigs of thyme in 1¾ cups of boiling water, covered and away from heat for half an hour. Strain and drink 3 to 4 cups a day.

TORMENTIL ∽ *Potentilla erecta.* **Parts used**: root. **Constituents**: Primarily, this herb contains catechol/tannins, as well as other constituents such as glycosides, resin, a red pigment, and gums. **Therapeutic benefits**: Tormentil has been used medicinally for sore throats and sore gums. It contains strong astringent properties. It has been utilized as a gargle. As an ointment, it can be useful for alleviating the pain and itch often associated with hemorrhoids. **Contraindications**: Because there is little recorded documentation regarding the safety of tormentil, it is suggested that it be avoided by pregnant or lactating women. **Recommendations**: Not to be swallowed.

VALERIAN ∽ *Valeriana officinalis* is a species belonging to the herb family known as Valerianaceae. **Parts used**: root. **Constituents**: glycosides, volatile oil, and several acids. The oil contains a completely different structure and chemical constituency. **Therapeutic benefits**: This herb is a perennial native to Europe and the United States. Valerian has been studied in Europe and Russia for its major constituents (the valepotriates).It is reported to produce sedative, anticonvulsive, blood pressure lowering, and tranquilizing effects. **Caution**: Prolonged or excessive use of valerian may result in side effects such as headaches, nervousness, irregular heartbeat, and insomnia. **Recommendations**: Half a dropper to a full dropperful of the liquid extract taken in water once daily (time of day to be determined by practitioner or herbalist, depending upon symptoms being addressed). This herb, as with most of the other herbs, should be taken in potencies and frequencies according to a qualified practitioner's recommendations.

WALNUT, BLACK ∽ *Juglans nigra*. **Parts used**: the fruit, leaves, and the inner bark. **Constituents**: Oils, tannins, quinones, and essential fatty acids. **Therapeutic benefits**: Herbalists have long recommended the fruit for adding strength and weight to individuals who suffer debility. The leaves and fruit are said to alleviate certain skin conditions, including eczema, herpes, and psoriasis. Constipation has been relieved through formulas that include the bark of the black walnut. Some herbalists have utilized black walnut leaves as a means of passing parasites; however, there are no studies to document black walnut's ability to accomplish this. Black walnut's inner bark quills are potent, yet safe laxatives. **Contraindications**: None have been documented. **Recommendations**: Depending upon desired effects, follow standardized potency recommendations for low doses, as provided for on manufacturer's label.

WALNUT, WHITE ∽ *Juglans cinerea* L., *butternut*. **Parts used**: dried inner bark and leaves. **Constituents**: quinones, tannins (the nuts contain essential fatty acids), and oils. **Therapeutic benefits**: The inner bark is used as a laxative, which is said to be safe to use in pregnancy. The leaves make a popular home remedy that is used in Europe for eyelid inflammation as well as for eczema. Walnut leaves possess an astringent property and have been utilized for expelling worms.

It is employed also as an antispasmodic, and the nut rind is an anti-inflammatory. **Contraindications**: There are no contraindications documented for white walnut. **Recommendations**: As with the recommendations provided for black walnut, follow the manufacturer's recommendations. Each manufacturer provides various potencies (most use the standardization provided by the British pharmacopeia), and there are always individual variables, such as age, sex, weight, and desired effect. However, always begin with the lowest dosage, and adjust accordingly.

WHITE WILLOW ∾ ***Salix alba*** is a species belonging to a family known as Salicaceae. **Parts used**: the bark. **Constituents**: glycosides, various phenolic glycosides, including salicin. Tannins are also present, as are other constituents such as catechins and flavonoids. **Therapeutic benefits**: White willow bark is the first discovered source of a substance known as salicin, and it was used in Europe to combat pain as far back as the mid-1700s. White willow bark was written about in ancient Egypt, and it was used for pain by the Greeks. After salicylic acid was extracted from white willow bark, it later led to the manufacture of what we now refer to as aspirin. Aspirin was proven effective against rheumatic fever, arthritis, gout, general pain, and neuralgia. White willow bark is converted through oxidation to salicylic acid in the body. **Contraindications**: None have been documented for white willow bark; however, it is not recommended that it be used in conjunction with any prescribed drugs for pain, without first obtaining medical approval. **Recommendations**: As with most of the recommendations provided herein, low doses for a brief duration are best.

WILD YAM ∽ *Dioscorea villosa, Dioscorea mexicana* **(Mexican yam)**, *Dioscorea paniculata, colic root, rheumatism root* (including Chinese yam). **Parts used**: root. **Constituents**: Glycoside saponins. Derivatives of yam are converted to progesterone. Steroid drugs derived from yam include corticosteroids, androgens, and estrogens. Wild yam root reputedly is said to lower blood cholesterol and blood pressure. The most widely prescribed birth control pill in the world, Desogen®, is made from the wild yam. **Contraindications**: None have been reported for wild yam root. **Recommendations**: Wild yam is mostly used by menopausal women for easing the symptoms often associated with menopause. It is available in a cream. Half of a tablespoon rubbed nightly on any area of the skin is usually sufficient to reverse many of the annoying symptoms that often result in response to hormonal fluctuations during menopause. Supervision by a physician or naturopath is advised. Pregnant or nursing women should consult a health practitioner before using yam products because of their hormone content.

WORMWOOD ∽ *Artemisia absinthium, absinthe.* **Parts used**: leaf, flowering tops. **Constituents**: volatile oil, carotene, tannins, and vitamin C. **Therapeutic benefits**: The azulenes in the plant are anti-inflammatory and reduce fevers. The leaf and its flowering top in an infusion is a tonic for the digestive system, liver, gallbladder, and the blood. Wormwood is a perennial herb native to the Ural Mountains, taken to Egypt very early in recorded history and listed as useful for treating headaches, and for eliminating pinworms, which is how it got its name! In Europe and North America, it was believed to combat poisoning, and it has been recommended for insomnia, certain types of liver problems, indigestion, and inflammation. **Caution**: In excessive potencies, or frequent doses, wormwood can become a narcotic poison. Symptoms of wormwood poisoning include headache, trembling, and convulsions. Ingestion of the volatile oil or the liqueur (absinthe) distilled from wormwood may cause gastrointestinal symptoms, nervousness, stupor, coma, and death. The use of wormwood has been banned in many nations. **Contraindications**: Avoid the volatile oil of wormwood, and the liquor absinthe. **Recommendations**: A diluted herbal tea of the dried herb for brief and temporary use should be safe; however, as with most of the herbs, supervision by a qualified practitioner is advised. To be avoided by pregnant or lactating women due to insufficient available information regarding the safety of wormwood.

YAM ∾ *See* **Wild yam**.

YARROW ∾ *Achillea millefolium* is a species of the Compositae family of herbs. **Parts used**: flowering tops, serrated leaf, dried stems and roots. (The whole herb.) **Constituents**: The flowerhead contains acids, amino acids, alkaloids, fatty acids, flavonoids, volatile oils, and several other constituents in lesser concentrations. This herb has been fully reviewed. **Therapeutic benefits**: The peppery leaf can be cut up and added to dishes, such as salads. The flowers have been employed for flavoring liqueurs, and the flowering tops are utilized as a digestive aid and as a mild diuretic. The flowering tops were also used for treating high blood pressure. Native Americans used a root concoction for maintaining muscle strength. **Caution**: Yarrow has been suspected of being a photosensitizer (may cause people to be sensitive to sunlight), although extracts have been reported to lack phototoxicity and to be devoid of properties known for their photosensitizing effects. **Contraindications**: In sensitive individuals, yarrow may trigger an allergic response, especially for those who are hypersensitive to other members of the Asteraceae/Compositae family. High doses may interfere with existing therapies for high or low blood pressure or blood-thinning. Yarrow is associated with sedative and diuretic effects. It should be avoided by pregnant or lactating women. **Recommendations**: Studies of yarrow show both potentially positive and negative effects. Therefore, the best recommendation offered at this time is to use it occasionally in an herbal tea form and in conjunction with several of the other proven herbs.

YELLOW DOCK ∽ *Rumex crispus, curled dock, out-sting, Rumex acetosa, sorrel, broad-leaf sorrel.* A species belong to the family Polygonaceae. **Parts used**: root. **Constituents**: anthraquinones, tannins, acids, and a volatile oil. **Therapeutic benefits**: This herb can be found commonly growing along the roadside, or upon a wasteland. In Europe, it was once used as the preferred remedy for treating anemia and scurvy. Yellow dock has a high iron content. Coughs and irritated throats have been relieved through a tincture that was prepared and recommended by many herbalists. Other disorders that responded well to yellow dock include skin diseases, arthritis, and gallbladder problems. **Caution**: The leaves contain oxalic acid, which can be toxic!

Contraindications: None have been documented or are available at this writing. **Recommendations**: As a nutrient, two capsules taken once daily on a full stomach is the usual recommendation.

YUCCA ∽ *Yucca liliaceae*, **Y.** *alorfolia*, **Spanish bayonet**, **Y.** *filamentosa*, all part of the family Agavaceae. **Parts used**: the entire plant. The leaves are used to weave baskets and leaf fibers are used for rope. Yucca is also a food source of Native Americans. The flower stalks are eaten when fully grown but before the buds have opened up. The fruit is eaten raw. **Constituents**: terpenoids, saponins, and several other constituents. **Therapeutic benefits**: The southwestern American Indians used this herb for hundreds of years to treat pain and inflammation of arthritis and rheumatism. **Caution**: Yucca can occasionally be purgative (causing bowel movement) and cause intestinal cramping. **Contraindications**: Prolonged use of yucca may inhibit the assimilation of fat-soluble vitamins (including A, D, E, and K). **Recommendations**: Low doses of yucca (either in the form of a juice or in capsule form) for brief and/or intermittent periods.

Glossary

Acidophilus: Bacteria that are able to ferment milk and that occur naturally as part of intestinal flora.

ADI: Acceptable Daily Intake.

AIDS: Acquired Immunodeficiency Syndrome.

Analgesic: A substance or drug used to relieve pain.

Anemia: A condition characterized by insufficient red blood cells.

Antagonist: An agent that acts in opposition to, or counteracts, another substance.

Antibiotic: A substance with the ability to destroy or inhibit a microorganism.

Antioxidant: A substance that prevents oxidation and can protect against free radicals.

Arteriosclerosis: Abnormal thickening and hardening of arterial (artery) walls, which results in loss of elasticity.

Atherosclerosis: Thickening and hardening of the walls of arteries due to abnormal fatty deposits.

Carcinogen: A cancer-causing substance.

Catabolism: Changes in cells that result in the breakdown of cell material, which yields energy and waste.

Catalyst: A substance that modifies or speeds up the rate of a biological reaction without being used up or changed.

Chelation: A process that changes minerals into easily digested forms.

Chelator: A chemical compound that binds with a metal and is used for removing lead or other metals from the body.

Cholesterol: A lipid present in cells that is used to regulate certain bodily functions. As a constituent of LDL it may cause arteriosclerosis.

Citric acid cycle: Bodily process in which a series of cellular chemical reactions occur, during which molecules are oxidized and energy is released. Also known as the Kreb's cycle.

Collagen: A fibrous protein that is the main constituent of the body's connective tissue, such as ligaments and tendons.

Deficiency: A shortage or inadequacy of a substance or nutrient that is necessary for health.

Diuretic: A substance that increases urine flow from the body.

DNA: Deoxyribonucleic acid. A chain of nucleic acid molecules in chromosomes that carry the genetic code, or blueprint, of cells.

Edema: The excessive accumulation of tissue fluid.

EFAs: Essential Fatty Acids. Unsaturated fatty acids (such as linoeic and linolenic acids) that the body requires for important functions and which the body itself cannot make and therefore must be obtained from dietary sources.

Enzymes: Complex proteins found in cells and necessary for accomplishing chemical changes in the body, such as breaking down food in the process of digestion or processing glucose (blood sugar) to create energy.

EPA: Environmental Protection Agency.

Extract: A product (an essence or concentrate) prepared by chemically or physically withdrawing the essential constituents (of a plant, for example).

Fatty acids: Organic substances that serve as building blocks for fat molecules.

FDA: Food and Drug Administration; a branch of the U.S. government's Department of Health and Human Services that is responsible for reviewing data required to establish safety, effectiveness, and proper labeling and manufacturing processes for all non-prescription medications prior to marketing. The FDA also regulated foods, cosmetics, and medical devices.

Fiber: Indigestible constituents in human food (roughage, bulk) that stimulate the digestive tract and benefit elimination.

Free radicals: Highly reactive chemical byproducts of cells that are generated by the body's normal chemical process of oxidation. Unwanted tissue changes, such as cancers, may result if free radicals are left unchecked.

GRAS: Generally Recognized As Safe; an FDA designation that covers substances added to food.

GTF: Glucose Tolerance Factor.

HDL: High-density lipoprotein; a cholesterol-poor and protein-rich constituent of blood plasma associated with a reduced risk of atherosclerosis; "good" cholesterol.

HIV: Human Immunodeficiency Virus; a virus that destroys certain cells of the immune system.

Hydrogenation: A process whereby an unsaturated oil, for example, is exposed to extremely high temperatures, breaking down the oil's molecular structure and changing it into a saturated(hydroginated) oil, which is solid at room temperature. Hydrogination of oils destroys the essential fatty acids while releasing free radicals.

Hydrolysis: A chemical process of decomposition.

IARC: International Agency for Research on Cancer.

Infusion: A product obtained by infusing; that is, steeping in water without boiling in order to extract the beneficial essences or constituents (of an herb, for example).

IU: International Units; a certain agreed-upon international standard for a quantity of a substance (such as a vitamin) that produces a particular biological effect.

Kreb's cycle: *See* citric acid cycle.

Lactating: Producing (breast) milk.

LDL: A cholesterol-rich, protein poor constituent of blood plasma that is correlated with an increased risk of atherosclerosis; "bad" cholesterol.

Lipids: Fats or fatty substances in the blood, such as cholesterol and triglycerides, that usually have fatty acids in their molecular structure.

MAO inhibitors: Monoamine Oxidase inhibitors; antidepressants that promote an increase in certain chemical messengers (neurotransmitters) in the brain.

Mutagen: An external agent that can increase the rate of mutation, or permanent change in the hereditary material of cells. Some chemicals, viruses, and forms of radiation can act as mutagens.

Myelin: Fatty material that forms the sheath-like coverings that enclose the body's nerve fibers.

Naturopathy: The system of disease treatment that emphasizes the use of natural agents, rather than drugs or surgery.

NRC: Natural Research Council; a U.S. government organization that sets minimal guidelines for the dietary intake of essential vitamins, minerals, and nutrients.

Oxidation: The union of a substance with oxygen, which yields a change in the structure of molecules. In cells, oxidation, often activated by enzymes, results in the release of chemical energy, used to promote cellular or chemical changes (metabolism).

PAFA: Priority Based Assessment of Additives; the FDA data bank that can provide consumer information on additives to food, cosmetics, and drugs.

Photosensitivity: Sensitivity or an adverse reaction to sunlight.

PPM: Parts per million.

Prostaglandins: A group of hormone-like substances, present in a wide variety of tissues and bodily fluids, that influence respiration, blood pressure, gastrointestinal secretions, and the actions of the reproductive system; they are also involved in the production of pain and inflammation in such diseases as arthritis. Prostaglandins have been used in the treatment of such disorders as asthma, ulcers, and high blood pressure.

RDA: Recommended Dietary Allowances (or Recommended Daily Allowances) of nutrients as determined by the National Research Council.

RNA: Ribonucleic acid; a substance that is involved in the control of the chemical activity of cells.

Saturated fats: Fats that are usually solid at room temperature, often found in foods from animal sources, which tend to raise blood cholesterol levels ("bad" fats).

Steroids: Sex hormones and hormones of the adrenal cortex;or cortisone-like medications prescribed when the adrenal glands do not produce sufficient hormones. Steroids have been used to treat inflammations and allergic reactions.

Synergist: Something (such as a chemical) that enhances the effectiveness of another agent.

Synthetic: Produced artificially, through chemical synthesis, rather than obtained from natural sources.

Therapeutic: Assisting in or providing a cure; medicinal.

Tincture: A solution of a medicinal substance, usually in an alcoholic base or carrier.

Toxicity: The condition or quality of being harmful or poisonous.

Triglycerides: Fatty substances in the blood.

Unsaturated fats: Fats that are usually liquid at room temperature, obtained most often from vegetable sources, which tend to lower blood cholesterol ("good" fats).

Volatile oils: Oils that vaporize readily (essential oils).

WHO: World Health Organization.

Consumer Resources

Questions regarding ingredients of processed foods, including reports of adverse reactions, should be directed to:

Office of Consumer Affairs, Food and Drug Administration, HFE-88, 5600 Fishers Lane, Rockville, MD 20857.

or

FDA in Washington, D.C. ("Hotline" open to consumer inquiries 10 a.m. to 4 p.m. Eastern Standard Time) 1-800-332-4010.

Questions or reports of contamination of meat and poultry should be directed to the **U.S. Department of Agriculture's Meat and Poultry Hotline** (10 a.m. to 4 p.m. Eastern Standard Time): 1-800-535-4555.

Also available to answer questions: **World Health Organization** (weekdays 8:30 a.m. and 4:30 p.m. Eastern Standard Time): 1-202-974-3000.

References

Bremness, Lesley, *Herbs*. New York: Dorling Kindersley, 1994.

Chaitow, Leon. *Thorsons' Guide to Amino Acids*. London: Thorsons/Harper-Collins, 1991.

Heinerman, John. *Heinerman's Encyclopedia of Fruits, Vegetables and Herbs*. West Nyack, N. Y.: Parker, 1988.

Lawless, Julia. *Illustrated Encyclopedia of Essential Oils*. New York: Penguin, 1995.

Mabey, Richard. *The Complete New Herbal*. New York: Penguin, 1991.

Newall, Carol A., Linda A. Anderson, and J. David Phillipson. *Herbal Medicines*. London: Pharmaceutical Press, 1996.

Ody, Penelope. *The Complete Medicinal Herbal*. New York: Dorling Kindersley, 1993.

Shook, Edward E. *Advanced Treatise on Herbology*. Pomeroy, Washington: Health Research, 1993; Brooklyn, N.Y.: Revisionist Press, 1991.

Winter, Ruth. *A Consumer's Dictionary of Medicines*. New York: Crown, 1993.

Additional Reading

Bricklin, Mark, and the editors of *Prevention Magazine*. *Nutrition Advisor*. Emmaus, Penn.: Rodale Press, 1993.

Fisher, Lynn. *Fat-Free Cooking*. Emmaus, Penn.: Rodale Press, 1997.

Grieve, M. *A Modern Herbal*. Edited by C. F. Leyel. New York: Random Century Group, 1994.

Harte, John, Cheryl Holden, Richard Schneider, and Christine Shirley. *Toxics A to Z*. Berkeley: University of California Press, 1991.

Margen, Sheldon. *The Wellness Encyclopedia of Food and Nutrition* and the Editors of the University of California at Berkeley *Wellness Letter*. New York: REBUS/Random House, 1992.

Schlesinger, Sara. *500 More Fat-Free Recipes*. New York: Villard Books /Random House, 1998.

Steinman, David, and Michael Wisner. *Living Healthier in a Toxic World*. New York: Perigee/Berkley Publishing Group, 1996.

Index

Personal Notes